Forest Policy and Ecological Change
Hyderabad State in Colonial India

S. Abdul Thaha

FOUNDATION® BOOKS

Delhi • Bangalore • Mumbai • Kolkata • Chennai • Hyderabad • Pune

Published by
Cambridge University Press India Pvt. Ltd.
under the imprint of Foundation Books
Cambridge House, 4381/4 Ansari Road, Daryaganj,
New Delhi 110 002

Cambridge University Press India Pvt. Ltd.
C-22, C-Block, Brigade M.M., K.R. Road, Jayanagar,
Bangalore 560 070
Plot No. 80, Service Industries, Shirvane, Sector-1, Nerul,
Navi Mumbai 400 706
10 Raja Subodh Mullick Square, 2nd Floor, **Kolkata** 700 013
21/1 (New No. 49), 1st Floor, Model School Road, Thousand
Lights, **Chennai** 600 006
House No. 3-5-874/6/4, (Near Apollo Hospital), Hyderguda,
Hyderabad 500 029
Agarwal Pride, 'A' Wing, 1308 Kasba Peth, Near Surya Hospital,
Pune 411 011

First Published 2009

ISBN 978-81-7596-632-1

Typeset at Laser Print Craft, New Delhi.

Published by Manas Saikia for Cambridge University Press India Pvt. Ltd. and printed at Sanat Printers, Kundli, Haryana.

Contents

‘Forest is a peculiar organism of unlimited kindness and benevolence that makes no demand for its sustenance and extends protection to all beings, offering shade even to the axe-man who destroys it….’

– Gautam Buddha

Preface

The management of the natural resources is important for the welfare of any country. In particularly agrarian societies, natural resources like land, water and forests played a significant role. Of late, the environmental degradation and destruction of natural resources like forests have attracted the attention of the scholars across the globe. The scarcity of the natural resources worldwide, which generated many conflicts in recent decades, has affected the quality of human life as well as the natural environment. This deterioration is not a recent phenomenon. It is a process evolved in history: in the forest policies and the economies that generated such policies. The study of these aspects is called environmental history, 'whose principal goal has become one of deepening our understanding of how humans have been affected by their natural environment through time, and, conversely, how they have affected the environment and with what result'.

In India too, whose economy is predominantly agrarian, people are dependent on its natural resources for their livelihood and sustenance. The management and utilization of natural resources like forests, land and water is crucial if viewed from the historical perspective. It has been observed by the historians that from nineteenth century onwards the management of natural resources have been increasingly brought under the control of the State. Though the government intervention was not unknown in the pre-British period, the major difference between the pre-British times and British period lies in the scale and magnitude of this intervention and the everlasting impact it has had on the society. Many historians have noticed the significance of the State as a leading, often an influential factor in shaping the policy towards the management and exploitation of natural resources in India. Many studies show that the State started dominating nature and utilized it to the maximum extent with its remarkable interventionist policies since the middle of nineteenth century. During this time various problems associated with forests, water and irrigation, fisheries, famines, drought and epidemics have become the centre of attention. This should also be noted that the formation of state Forest Departments, imposition of several Forest Acts and application of its policies by way of day-to-day administration have had 'an unprecedented and ever-lasting impact on the Indian society'.

With the growing interest in the regional studies, there are several valuable studies on the environmental history of specific regions in India. However, most of these scholarly works have been conducted during the last three decades on the identifiable ecological zones of the British India, like the Himalayan region, Punjab, Central Provinces, Bombay and Madras Presidencies. The native states which constituted one third of the sub-continent under the British government, have not received much attention from the historians. One of such native states is the Nizam's Dominions or popularly called as the Hyderabad State, then the biggest princely State in India.

The present book sets out to examine the nature of forest policy and its effect on natural resources, ecology, etc. in the Hyderabad State, during 1867–1948. The choice of Hyderabad State rests on the fact that, though it was an independent State, in many aspects it resembled a British territory, particularly in the field of state forestry. Concentrating on 71 years period, i.e., from the date of establishment of the Forest Department in the State in 1867 to the amalgamation of the State into the Indian Union in 1948, this work examines forest polices of the State and the changing relationships between the state Forest Department and the people dependent on forest resources.

This phase in the history of Hyderabad State was particularly important; as in course of time the forests witnessed a transition from being a property of people to being the property of the State. This had far-reaching implications for the organization of Forest Department and its relationship with the peasants and tribal populations. The important changes introduced into the management of forests in the Hyderabad State were: opening of Forest Departments, taking over of forests by the government, denial of traditional forest rights of peasants and tribal communities, restrictions on the mobility of common people in forest areas, introduction of commercial forestry and negligence to tackle the effects of deforestation. In fact all the above changes were introduced by the British in their directly ruled areas which were adopted by the Nizam's government in Hyderabad State. The Forest Department in the Hyderabad State was itself established on the model of the British Forest Department in United Kingdom which in turn had been influenced by the European ideas and practices. The present study is aimed at highlighting the history of forest policies and the concomitant changes that took place in the backdrop of British colonial influence in the State. Thus, the study of forestry in the Hyderabad State would help understand the regional variations with the forested areas of other states.

I wish to thank State Archives of Andhra Pradesh, Tamil Nadu and National Archives of India, New Delhi for providing access to their archival sources. I also thank the Librarians and the staff of the Libraries of University of Hyderabad, Osmania University, State Central Library, AP Forest Department, Centre for Economic and Social Studies, Telugu University, Sundaraiah Vignana Kendram, Abul Kalam Azad Oriental Research Institute (Hyderabad) and Jawaharlal Nehru University (New Delhi) for making available the required source material for the study.

My sincere thanks go to Professor Atlury Murali, Head of the Department of History, University of Hyderabad, for his constant help, encouragement and guidance during the research work. Support and cooperation of other faculty members of the Centre for Regional Studies and Department of History, University of Hyderabad are also thankfully acknowledged.

In the course of research, I have come across a number of academicians and scholars with whom I could have fruitful academic interaction. I would like to name and specially thank Professor Laxman Satya (Lock Haven University, USA) and Dr Neena Rao (Hyderabad) for their constant support, valuable advice and encouragement. Kind help of my colleagues and friends Dr D. Venkateshwarlu, Dr Ch. Srinivas, Dr Mustafa, Dr Saidulu, Dr P. H. Mohammad and Mr Sandeep is also thankfully acknowledged. I warmly thank all those who helped me complete this monograph in one way or the other.

Finally, I must record my sense of obligation and sincere respects to my parents and other family members to pursue my interest. Last but not least, I thank my wife Shahnaz for being a source of inspiration and encouragement.

S. Abdul Thaha

Hyderabad State in Pre-independence India

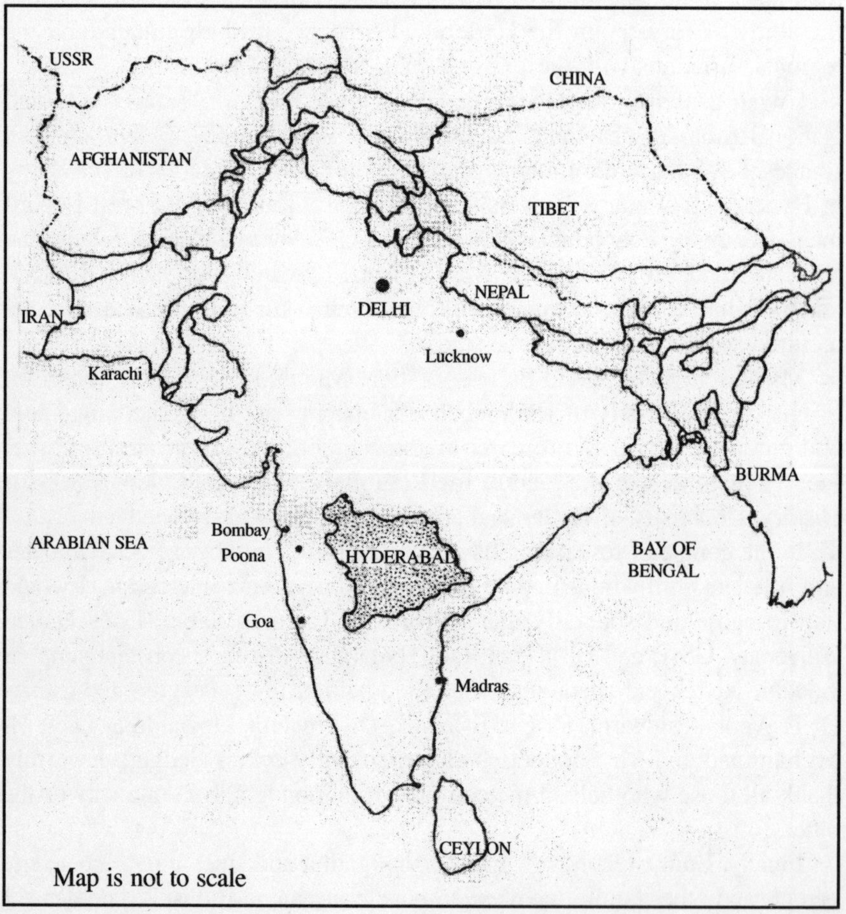

Political Map of Hyderabad State

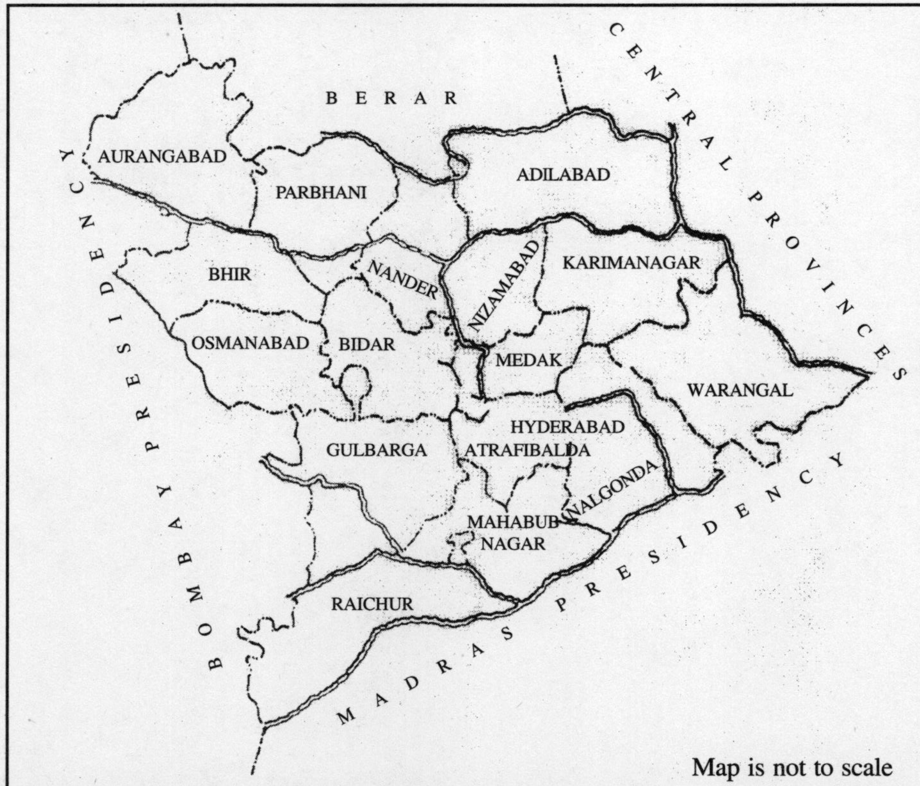

CHAPTER 1

Introduction

The management of the natural resources is important for the welfare of any country. In a predominantly rural country like India, where people are undoubtedly dependent upon their natural resource base for their sustenance, the management and utilization of natural resources like forests, land and water is crucial. As the increase in population and concomitant demands grow over control, use of natural resources makes resource-use patterns change. Though the government intervention was not unknown in the pre-British period, particularly from the nineteenth century onwards the natural resources have increasingly come under the control of the state.

Many historians have noticed the significance of the state as a leading, often the influential factor in shaping the policy towards the management of ecological resources of India. Particularly, the second half of the nineteenth century is considered to be a period of 'remarkable interventionist policies; an age of high imperialism and supreme confidence in the capacity of science and technology to control nature and utilise it to the maximum extent'.[1] During this period various problems associated with forests, like irrigation, famines, drought and epidemics became the centre of attention. Concern is expressed over the above said environmental issues by experts in

---

1 For details at all India level, see David Arnold and Ramachandra Guha (eds.), *Nature, Culture, Imperialism: Essays on the Environmental History of South Asia,* Oxford University Press, New Delhi, 1995; Ramachandra Guha, 'An Early Environmental Debate: The Making of the 1878 Forest Act', in *IESHR,* vol. 27, no. 1, 1990, pp. 64–84; 'Forestry in British and Post-British India: A historical analysis', in *EPW,* vol. 18, no. 44 and 46, Oct. 29 and No. 12, 1983, pp. 1882–1896 and 1940–1946.

various fields. These disturbances in the ecosystem have led several scholars to reconstruct history, using insights derived from recent debates on environmental issues. The historical study of forestry and forest policies is relatively a recent phenomenon and one of the positive results of such concern has been the emergence of environmental history as a distinct field in the historical writing.

One of the most important recent developments in the field of environmental history since 1970–80s has been the growing interest shown by historians in the process of ecological change. Since then, the socio-ecological research has been rapidly gathering pace. In the context of ecologically oriented history of India, it starts with the period of British colonial rule which has been identified as a crucial ecological watershed, in the sense that it brought with it novel techniques of managing and controlling nature by enacting laws and procedures. The main emphasis of research in this field is on the historical process of the state forestry and legislation, the traditionally held rights of forest communities and their curtailment by the state. The techniques and management tactics adopted by the Forest Department in order to achieve its goals can be traced in its socio-historical context. The focus is also on the conflicts generated by technical change and by the attempts of the government to integrate the forests into the larger colonial economy.

With the growing interest in regional studies, of late, substantial research has been carried out on the environmental history of specific regions of British India, viz, the Himalayan region, Punjab, Central Provinces, Western Ghats and Madras Presidency. Despite significant advancement in the field of environmental history in the British territories, the native states which constituted one third of the Indian subcontinent, have not been paid due attention from the historians and have hardly been studied from ecological point of view. One of such native states is the Nizam's Dominions or the Hyderabad State, then, the largest princely State in India could not attract their attention. The purpose of the present monograph is to make an attempt to fill in this gap. It tries to contribute to other regional studies to construct a comprehensive environmental history of India.

The Study Area and Objectives

The present work sets out to examine the nature of forest policy and its effect on natural resources, ecology, etc. in one of the major Indian native states, the Hyderabad State, during 1867–1948. The erstwhile Hyderabad State, which was endowed with rich biological diversity besides varied and rich cultural spectra, offers an excellent scope for this study. The choice of

Hyderabad State rests on the fact that, though it was an independent state, in many aspects it resembled a British province, particularly in the field of state forestry as it was largely manned and influenced by the British personnel.

The present study is aimed at highlighting the history of forest policies and the concomitant changes that were brought in the backdrop of British colonial influence in the state. It has attempted to present a comprehensive picture of the history of forests and forest policies and management in the Hyderabad State covering the period 1867 to 1948. The time framework is specially chosen as it represents a period of great political and economic significance that is manifested in the imposition of state authority over the forests which was hitherto not known earlier. This study also analysed the underlying commercial aims of the state forestry and examined to what extent social and environmental considerations were incorporated into the forest policies.

The monograph has attempted to study the traditional management of forests prior to 1867 and also focused on the demographic description of forests and the people/tribals living in and around the forest areas. It has also analyzed the history of forest administration, evolution of forest policies and management in Hyderabad State. Various stages of the forest policy of the state between 1867 and 1948 are also taken into account while preparing the report. The present work has also examined the socioeconomic relationship between the tribal communities, forest resources and the forest policy of the state.

Another objective is to find out the relationship between forestry and agriculture. Prior to the establishment of Forest Department in the Hyderabad State, forests were viewed as an obstacle for the expansion of agriculture. With the state committed to agrarian expansion as a major source of revenue, the Hyderabad State allowed destruction of the forests and gave incentives to peasants in this matter, such as giving *pattas,* etc. in forest areas. The study focuses on the relationship between the two contradictory policies of extension of forest and cultivable areas by respective departments and the efforts made to resolve this problem in the later years. It also focuses on famines and scarcity in the state and their influence on forests.

The entire argument would be discussed in detail by making an in-depth critical analysis of the evolution and development of the forest policies during 1867–1948 which had long-lasting effects on the environment, management and commercialization of forests and socioeconomic relationship of peasants and tribal communities with the forests. It has tried to study forest policies and the links between the competing systems of the forest use and management, ecological changes and agrarian/tribal protests. Besides this, the impact of state's interventionist environmental policies on forests and on agrarian economy and peasant protest is also analysed. Finally, the study has tried to explain the changes in the forestry, ecology and changes in the forest policies of the state and the British colonial influence in this regard.

Methodology and Sources

The study begins with the premise that the establishment of Forest Department brought changes in the forests which were shaped in such a way so as to suit the political and economic needs of the state. It has also taken into account that the state policy was concentrated more to maximize the revenue with minimum expenditure. The attitude of the Forest Department was highly influenced by the general trend by neglecting the feelings of the subjects.

For the purpose of studying the above said aspects, the monograph relies on primary as well as secondary sources, which are numerous and varied. They include from private papers of nobles of the state to the government reports and records. While referring to the government records, a special care has been taken as almost all government records tried to justify their activities and representing themselves as serving the people, while the reality was largely different. Hence, before the data is used one is reminded of such problems associated with the government records.

The data for the study undertaken is found in different source materials. For example, reports of the Forest Department speak about the existence of forests and provide deeper insights to the functioning of the department from the government perspective. For the profile of the state forest area, legislation, acts, etc. the published government sources are used extensively. For a general understanding, gazetteers/manuals have been used. As the study seeks to analyse real nature of forest policies, and motto of the state in this regard, the regional language sources like people's views published in newspapers, magazines and proceedings of organizations like Andhra Maha Sabha, Communist Party literature, etc., are compared. Some of the books were written by contemporary writers at different times, capturing the spirit of the times. While other modern works by scholars, who have carried out research on different aspects of ecosystems, have been consulted to strengthen the arguments and presented in a theoretical framework.

However, all the available works give us only a broad picture of the forest area and the geographical description of the country but not any comprehensive and analytical account of state forestry, its policies, administration and effects of the forest policies in the historical context. Further, compared to British India, the working of state forestry and forest policies in the Hyderabad State during the late nineteenth and first half of the twentieth centuries has not been studied from ecological point of view. Colonial state forestry, control over forests by the state, erosion of traditional rights of forest dwellers and peasants, etc. have been studied for British ruled areas, but were completely absent in the studies on the Hyderabad State. Since there is no systematic or exclusive study dedicated to environmental history, the present study is a modest attempt to fill this lacuna in the understanding of the state forestry, forest policies and consequent ecological change in the Hyderabad State.

Geographical Conditions

Physical Features

The Hyderabad State was in the centre of Deccan plateau situated between 15° 10' and 20° 40' north latitude and 74° 90' and 80° 35' east longitude. It was surrounded by the Central Provinces in the North, the Andhra area of the undivided Madras Presidency in the South and the East, and the Bombay Presidency on the Western side.[1] The Hyderabad State, which was the largest among the native states of pre-independence period, covered an area of 82,698 sq. miles.[2] It had a population of 98,45,594 in 1881 with a density of 119 persons per sq. mile and raised to 186.55 lakhs in 1951 with a density of 227 persons per sq. mile.

The average altitude over the sea level was 1250 feet and certain summits were even well above 2500 to 3000 feet. The state was traversed by several hill ranges like the Balaghat range, the Sahyadri Hills and the Jalna Hills in addition to numerous clusters of hills of minor importance.[3] There were, however, certain undenuded portions of a series of flat-topped plateau (locally known as hills) belonging to the great tableland of peninsular India, now forming watersheds between the different drainage areas. There was one main watershed traversing the state from north-west to south-east dividing the country into two nearly equal basins, which were drained off, by the

---

1 *Imperial Gazetteer of India, 1909, Provincial Series, Hyderabad State,* Superintendent of Government Press, Calcutta, 1909, p. 1.

2 *Census of India, 1911,* vol. XIX, Hyderabad State, Part 1, Times of India Press, Bombay, 1913, p. 1.

3 *Report of the Administration of HEH the Nizam's Dominions for 1303 Fasli (1893–94),* Times of India Press, Bombay, 1895, pp. 2–3.

Godavari and the Krishna rivers.[4] The important rivers were the Godavari, the Krishna, the Penganga and the Tungabhadra besides more than 45 streams which could be harnessed for irrigation.

The entire state was divided into two large equal divisions geographically and ethnically known as Marathwada and Telangana and a small division of Karnataka.[5] These divisions were separated from each other by the Manjira and the Godavari rivers. The portion to the north-west formed the trappean region and the granitic and calcareous (limestone) region of the south-west. Ethnically the trappean region was inhabited by the Maratha race and the granite region by those of Telugu and Kanarese speaking population of the south-east. The difference between these two natural divisions in respect of soils, vegetation and hill features had its direct and profound influence on cropping pattern, type of cultivation, habits and the economic well-being of the people of these regions.

The main river systems of the Hyderabad State belonged either to the Godavari river in the north or to the Krishna river in the south. The Godavari is the most important river of the state with a total length of about 700 miles running through the districts of Aurangabad, Parbhani, Nanded, Nizamabad, Adilabad, Karimnagar, Warangal and Khammam in the state and the enters the Godavari district of the Madras Presidency. The Krishna is next in importance to the Godavari with a total length of 400 miles, flowing through the districts of Raichur, Mahaboobnagar, Nalgonda and Warangal.[6]

There were no natural lakes in the Hyderabad State. The undulating nature of the country helped in the construction of artificial tanks or dams across the valleys of small rivulets and streams, to intercept water during the rainy season for irrigation purposes. The total number of tanks, large and small, in the state was 20,476.[7] In fact, such artificial tanks are particular to the granite country (Telangana) and whenever groups of granite hills occur, tanks are sure to be found associated with them. They are not generally found in the trap regions such as Marathwada, as the soil is too porous over there and the *bunds* thrown across become much cracked and fissured in the hot season, so that they are easily breached on the burst of rains.[8]

---

4 *Administrative Report of the HEH the Nizam's Dominions, 1331 Fasli (1921–22)*, Companion volume, Government Central Press, Hyderabad, 1925, p. 1.

5 *Imperial Gazetteer*, pp. 1–2.

6 Mahdi Ali, *Hyderabad Affairs*, vol. vii, Bombay Exchange Press, Bombay, 1879, pp. 30–31.

7 HEH the Nizam's Dominions, *General Administration Report*, 1892–93, p. 6.

8 Ibid. Also in the Department of Agriculture, HEH the Nizam's Government, File No. 35, Serial No. 7 of 1357 Fasli (1948), Andhra Pradesh State Archives (APSA), Hyderabad.

A large area of the state consisted of rocky soil rich in mineral resources such as coal, iron, limestone, mica, graphite and gold. Iron ore deposits of varying quality were widely distributed in the Godavari and Wardha valleys. Hematite was found in the Krishna and Tungabhadra valleys. Even diamonds in the Kurnool series and gold in Raichur were mined. The only mineral of great economic value that was being worked in the state was coal. The coal mines were situated at Singareni, Sasti and Paoni. Some other minerals were also found in the state like mica, corundum and garnets in Warangal district, graphite in Karimnagar and limestone in Shahbad. The Shahbad stone was widely used for flooring and manufacture of cement.[9]

The region had the tropical climate and the mean temperature was about 27 degrees Celsius, the maximum in summer often rose to above 48 degrees Celsius in some low lying parts of Telangana.[10] It received an average rainfall of 33 inches in the East to about 29 inches in the west, mainly brought by the south-west monsoon according to the figures of rainfall from 1900–1940. The distribution of the rainfall is capricious. Some of the Telangana tracts in the north-east corner receive an average 40 to 50 inches of rain. While the Karnataka with an average rainfall of not exceeding 25 inches, suffers from frequent occurrence of drought.[11] The chief crops produced in the state were rice, wheat, cotton, oilseeds and millets. The region had 9,357 sq. miles of forest constituting 11.7 per cent of the total area.

Political and Administrative History

Many political and historical changes happened in the Hyderabad State after the decline of the Mughal Empire in the early part of the eighteenth century. The Qutb Shahis had ruled over the Deccan region before it was occupied by the Mughal Emperor Aurangzeb in 1687. In 1713, the Mughal Emperor Faruqsiyar appointed the Nizam ul mulk Asaf Jah I, an able General of Aurangzeb, as the Viceroy of the six Deccan provinces and *fouzdar* of the Karnatak. He was given the title of Nizam ul mulk Bahadur Fateh Jung, which had since become the hereditary title of the dynasty.[12] However, due to politics and factions at the imperial court, he resigned and marched down to the Deccan informing the Emperor that he felt his imperative duty to control the Marathas. Knowing the

---

9 *Souvenir of His Royal Highness the Prince of Wales' Visit, Hyderabad State,* Government Central Press, Hyderabad, 1922, p. 22.

10 *Some Economic Facts and Figures of HEH the Nizam's Dominions,* Bulletin No. 1, Government Central Press, Hyderabad, 1937, p. 2.

11 Ibid., p. 3.

12 *Administrative Report of 1303 Fasli (1892–93),* p. 9; Bharati Ray, *Hyderabad and British Paramountcy, 1858–1883,* Oxford University Press, New Delhi, 1988, p. 2.

intentions of the Nizam, the Emperor Muhammad Shah instructed Mubariz Khan, the Governor of Khandesh, to put down the Nizam. In a battle fought in 1724, the Nizam emerged victorious and established Asaf Jahi dynasty in the Deccan.[13] However, neither the Nizam I nor his immediate successors openly claimed independence of the Deccan from the Mughal Empire. They continued to pay allegiance to the Mughal Emperor and acknowledged formally that the Nizams administered government as a delegate of the Emperor.[14]

Due to political affairs in the Karnataka region, the Nizam I developed direct contacts with the French and the English who were fighting with each other for commercial supremacy in the Coramondal coast. Although the Nizam was in favour of the French, he did not involve seriously in the war between the foreign traders and allow them to exert any influence on the internal affairs of his state. The Nizam I, in the midst of the confusion that characterized the dissolution of the Mughal Empire, appeared as one of the most commanding figures of the time.[15] After the death of Nizam I in 1748, the period 1748–1761 was of intrigues among his aspiring sons to the throne and rivalries between the French and British in the state. In the dispute over succession, Nizam Ali Khan, the fourth son of Nizam I, ascended the throne in 1762 with the help of the British. Nizam II ruled for 42 years and signed a series of treaties with the British in which he assigned large territories in return for the defence of the state in times of wars. Gradually, the British strengthened their power and increased their interference in the Hyderabad State affairs. Treaties were concluded in the years 1766, 1792, 1798 (Subsidiary Alliance), 1799 and 1800 which were all made favourable towards the British. As a continuation of the Subsidiary Alliance, a treaty was concluded in 1800, which forbade the Nizam to negotiate with the foreign powers without consulting the East India Company, except in some personal matters.[16] With this treaty the Nizam's sovereignty was curtailed as

---

13 *Administrative Report of 1303 Fasli (1892–93),* p. 10; *Imperial Gazetteer of India,* p. 15.

14 *Administrative Report of 1331 Fasli (1921–22),* pp. 20–21.

15 P. Setu Madhavan, *18th Century Deccan,* pp. 61–63; and Syed Abid Hassan, *Whither Hyderabad,* Madras, 1935, p. 2. The early history of the Asaf Jahi dynasty was that of a struggle for existence against the Marathas and the Europeans. The years that followed the death of Nizam I were of uncertainty and confusion. Successors who were less discrete, weak in both diplomacy and military skill unlike Nizam I, started wars of succession among themselves, and were forced to bow to the French and the British. Consequently, they became embroiled in their Anglo-French wars in south India and worse still, gave the foreigners an opportunity first to intervene in and later to establish control over their states.

16 *Maharashtra State District Gazetteer,* Parbhani District, Government Press, Bombay, 1967, p. 68; C.U. Aitchison, *A Collection,* pp. 67–73.

his power to deal with the foreign countries was gone and the East India Company was left free to make peace or war in any part of India without consulting the Nizam. Thus Hyderabad, once an equal ally, became essentially a protected state. Henceforth, the chief interest of the political history of Hyderabad lies in the nature of the British control exercised over the state.[17]

During the rule of Sikandar Jah, the Nizam III (1803–1829), the British interfered a great deal in the internal administration of Hyderabad State. The State was already under the Company's control in all its external affairs, it would now be dominated internally by the Company as well. The British decided to establish its sway over the State through the Resident officer on the matters of the appointment of pro-British prime ministers and other administrative matters on the pretext to set right the administration in the state.[18] On the death of Nizam III in 1829, his elder son Nasiruddaula (Nizam IV) succeeded to the throne and ruled the state till 1857.

The period of the first four Nizams from 1713 to 1857, may be described as the era of confusion leading to financial instability of the State. This period witnessed the worst hit of famines and depredation. The sources of finance were limited and unorganized. Every department of the State became disorganized and the credit of the State became so bad that even bankers refused to grant loans. The feudal lords entrusted with the collection of revenues, exploited the opportunity and gave great trouble to the people. The land revenue formed the core of the state finance and the collection of which was the responsibility of the *taluqdars*. The *taluqdars* in turn assigned the duty of collecting the land revenue from the *ryots* to *tahsildars*. Ministers like Chandulal and Sirajulmulk tried to carry out several reforms and a system of revenue administration based on modern principles was evolved. The Hyderabad State was divided into 16 districts, each in charge of a *taluqdar* who was responsible for civil and judicial administration of his district.[19] Yet the financial crisis continued in the State and the government was overburdened with debts and the state treasury became empty. To meet the expenses even some of the jewels of the Nizam had to be mortgaged.[20]

---

17 Sarojini Regani, *Nizam–British Relations,* 1724–1857, Swarajya Printing Works, Secunderabad, 1963, p. 185.

18 Sarojini Regani, *Nizam–British Relations,* p. 187.

19 Raja Ali Khan, *Hyderabad 400 Years,* Zenith Services, Hyderabad, 1991, p. 97.

20 K. K. Mudiraj, *Pictorial Hyderabad,* vol. II, Chandrakanth Press, Hyderabad, 1934, p. 256.

This situation arose mainly because of the heavy expenses incurred in maintaining the Hyderabad Contingent in the State. Salarjung I took over the administration as a Minister of the Hyderabad State at a very critical period (in 1853) with limited finances and strained relations with the British. However, despite all these handicaps Salarjung I introduced several reforms and reorganized the administrative machinery, which resulted in improved finances of the State.[21]

Nizam IV was succeeded by his son Afzaluddaula, the fifth Nizam on 18 May 1857 and Indian mutiny started on 17 July 1857. This was a good opportunity for the Nizam to go against the British. During the mutiny the Governor of Bombay Presidency telegraphed to the British Resident at Hyderabad saying, 'If Nizam goes, all goes.'[22] Though some advisors of the Nizam urged him to raise against the British, but the Nizam listened to the advice of his minister Salarjung I and supported the British. After the storm of the mutiny had subsided, being pleased with the faithful attitude of the Nizam, the British forced him to sign some more new treaties. The year 1857 was an important year as the Mughal dynasty vanished from India and the Nizam ceased to be a Mughal *subedar* and later the State got free from any internal revolts.[23]

It was in the year 1858 that Salarjung I embarked his scheme of reforms in administration. Based on the British revenue administration, the State was divided into 5 divisions and 17 districts which were headed by *subedars* and *taluqdars* respectively who were paid in cash salaries.[24] In 1860, the Nizam alloted land to the Bombay–Madras railway line passing through the Hyderabad State in the Raichur doab which was later extended up to Wadi. A new line called Nizam State Railway was introduced to connect Hyderabad from Wadi. The introduction of a railway linked Hyderabad contributed to the development of economic activities in the State. During the reign of Afzaluddaula (Nizam V), the reforms carried out by Salarjung I included the establishment of a Government Central Treasury in 1855, Stamp Paper Office in 1856 and more importantly the establishment of the Forest Department in 1867. At the same time the Judicial, Public Works, Medical, Municipal,

---

21 For a contemporary account of the life and works of Salarjung I, see S. H. Bilgrami, *A Memoir of Sir Salar Jung I,* The Times of India Steam Press, Bombay, 1883.

22 St. Nihal Singh, *The Nizam and the British Empire,* (privately printed for the author, 1923), p. 1., quoted in Bharati Ray, *Hyderabad-British Paramountcy,* p. 10; and in Raza Ali Khan, *Hyderabad 400 Years.,* p. 99.

23 Raza Ali Khan, *Hyderabad 400 Years.,* p. 99.

24 *Revenue Administration Report 1914–15,* pp. 14–19.

Police and Education Departments were brought into proper organization. The system of assessment was thoroughly overhauled and the land revenue administration was reorganized on the lines of Bombay and other adjacent Presidencies.[25]

Nizam V was succeeded by his son, Mir Mahaboob Ali Khan in 1869. Since he was only three years old then, a regency was constituted under Salarjung I. This gave Salarjung a good opportunity to continue his reforms in the state administration till his death in 1883. As a result of his efforts, almost every department of the British administration was represented in the State and special attention was given to the improvement of the unbalanced finances.[26]

Mahaboob Ali Khan, the Nizam VI, after attaining majority in 1884 took the charge in his hands. During his rule, well-known ministers like Salarjung I and II, Sir Asman Jah, Sir Vicar Ul Umra and Maharaja Sir Kishen Pershad, were in charge of the government as the *diwans*. In 1892, a code known as the *Qanuncha-i-Mubarak* (the auspicious code) was issued for the guidance of the minister, and this was followed by the establishment of a council consisting of all the ministers of the State.[27] The administrative reforms of Mir Mahaboob Ali Khan included development of railways, revision of revenue settlements, establishment of cotton mills at Hyderabad, Gulbarga and Aurangabad for the benefit of cotton-growers. Large tracts of wasteland were brought under cultivation by undertaking new irrigation works; and with the restoration of old tanks economy was put on a sound footing with surplus budget. Education was also given special attention and a number of schools were established in all parts of the State. Police, Judiciary, Excise and Forest Departments were reorganized resulting in better revenue.[28]

The last member of this dynasty, Mir Osman Ali Khan, the Nizam VII, succeeded to the throne in 1911. During the reign of Nizam VII, some important measures were taken to improve administration, develop natural resources, establish cultural institutions and ameliorate the condition of people. In 1914 the First World War broke out and it had a profound effect on India, including the Hyderabad State. The Nizam rendered his assistance by

---

25 *Maharashtra State Gazetteer,* Nanded District, 1971, p. 82; and P. V. Kate, *Marathwada under.,* p. 19.

26 *Imperial Gazetteer of India,* p. 18–19 and Fathaulla Khan, *A History of Administrative Reforms in Hyderabad State,* pp. 65–66.

27 *Administration Report of 1331 Fasli (1921–22),* pp. 21–22 and *Imperial Gazetteer of India,* p. 19.

28 Maharashtra State Gazetteer, Nanded district, 1971, pp. 86–87.

providing his services to the British, and the British government conferred on him the titles of 'His Exalted Highness' (HEH) and 'Faithful Ally of the British Government' in 1918.

In the field of education the Osmania University was established in 1917 in Hyderabad, while intermediate colleges were set up at divisional headquarters from 1927. The Kazipet Ballarshah railway route and the railway line connecting Bidar and Purli were also opened. The Archaeological Department of Hyderabad State was constituted in 1914 to conserve the principal monuments of the State such as the Ajanta frescoes.[29] In 1922, judicial administration of the State was separated from its executive administration and Hyderabad was more or less a pioneer in this field. An industrial fund was opened with an initial capital of one crore rupees for the industrial development of the State. Large industries like Shahbad cement, Nizam sugar factory, Singareni collieries, Sirpur paper mills, etc. were also established during this time. The other big achievements of his rule were the construction of several big irrigational projects such as Nizamsagar, Alisagar and Manair. A dam was constructed in 1920 across the river Musi to avoid recurrence of floods and to provide drinking water for Hyderabad city at a place called Osman Sagar and later at Himayat Sagar.[30]

On the other side, the establishment of the Hyderabad State Congress, a branch of Indian National Congress, was opposed by the Nizam government who tried to suppress nationalist movements.[31] It also tried to curb the communist movement which was fighting against the evils of feudal lords and the suppressive nature of the State. With support of the nationalist organizations like Arya Samaj, Civil Liberties Union and the Hyderabad

---

[29] Ali Yavar Jung, *Hyderabad in Retrospect,* Bombay, 1949, pp. 12–13.

[30] B. K. Narayan, *Finances and Fiscal Policy,* p. 4, and Raza Ali Khan, *Hyderabad 400 Years,* p. 109.

[31] Government had decided to prohibit the formation of any organization 'so named or so constituted and the Hyderabad State Congress, is, if formed, declared to be an unlawful association under Public Safety Regulations.' See, N. Ramesan (ed.), *The Freedom Struggle in Hyderabad,* vol. IV, Andhra Pradesh State Committee appointed for the compilation of a history of the freedom struggle in Andhra Pradesh, Hyderabad, 1966, pp. 185–86. According to P. Sundaraiah, the Indian National Congress had adopted a policy of non-interference in the native states. Consistent with its bourgeois comprising theories of bringing a certain amount of mass pressure to get certain concessions without allowing the anger and surge of the masses to go beyond its control to revolutionary channels. The leadership of the Indian National Congress did not encourage the struggles of the people of the states against the native princes. For more details see, P. Sundaraiah, *Telangana People's Struggle and its Lessons,* Calcutta, 1972, p. 75.

State Congress, people began to express their resentment against the repressive policies of the State.[32]

During Sir Stafford Cripp's Mission to India in 1942, the Nizam entertained the plea of 'Independent Hyderabad'. Sir Stafford Cripps assured the Nizam's delegation that when British India would become independent, the native states would also be free to either join the Union of India or remain independent.[33] The Mountbatten Plan, while partitioning India and transferring power to the two Dominions, could not solve the problem of princely States. When the British officially left India on 15 August 1947, Hyderabad was an independent State and the Nizam decided to remain independent but it was not feasible for the Nizam to remain so under the prevailing circumstances and geographical reasons. All the efforts of moderators like Syed Mirza Ismail and Sir Sultan Ahmed to establish relations between the Indian Union and Hyderabad in consonance with the realities of the situation were opposed by the *Rajakars* (volunteers) under the leadership of Kasim Razvi.[34] The movement of the Hyderabad State Congress under the leadership of Swami Ramananda Tirtha to force Hyderabad to join Indian Union was strongly opposed by the Nizam's government.[35] The state of anarchy coupled with attempts of the Nizam to assert independence necessitated the intervention by the Indian Union through 'police action'. On 19 September 1948 Hyderabad was occupied by the Indian army and the Hyderabad State was forced to merge with the independent Union of India.[36] A military government was initially appointed to ensure security and was later followed by a civil government.[37] In November 1956 the states were reorganized on linguistic basis and the various regions of Hyderabad State were distributed among the states of Andhra Pradesh, Maharashtra and Karnataka.

---

32 Ramananda Tirtha, *Memoirs of Hyderabad Freedom Struggle,* Bombay, 1966, p. 189; and N. Ramesan, *The Freedom Struggle in Hyderabad,* vol. iv, Hyderabad, 1966, pp. 242–244.

33 N. Ramesan, *The Freedom Struggle*, pp. 226–238.

34 R. Tirtha, *Memoirs,* pp. 83 and 179; Laik Ali, *Tragedy of Hyderabad,* pp. 7 and 30.

35 *Karimnagar District Gazetteer,* pp. 63–64; Hyderabad State Congress, *The Nizam Disowned,* Hyderabad, no date, pp. 8–14.

36 V. H. Desai, *Vandemataram to Janaganamana: Saga of Hyderabad Freedom Struggle,* Bharatiya Vidya Bhavan, Bombay, 1990, p. 143.

37 B. K. Narayana, *Finances and Fiscal Policy of Hyderabad State (1900–1956),* Hyderabad, 1973, p. 4. N. Ramesan, *The Freedom Struggle,* p. 290.

CHAPTER 2

History of Forests

Classification of Forests

The types of forest vegetation in a locality depend upon the climate, soil and past treatment. Forest vegetation of the Hyderabad State was a part of the great deciduous belt occupying a large portion of peninsular India, stretching further north to the Central Provinces, eastward to Bihar, westward to the Bombay Presidency and in south up to the Madras Presidency. As per H. G. Champion and S. K. Seth's classification, the forests of the Hyderabad State region came under the group 'Southern Tropical Dry Deciduous Forests' and 'Southern Tropical Thorn Forests'.[1]

The main characteristics of the first type of forests are entirely deciduous or nearly so; top canopy rather light and rarely over 25 metres length, usually between 8 to 20 metres. Thorn forests are deciduous with low thorny trees having xerophytic features, canopy more or less broken and height is under 10 metres.[2] It includes forests consisting of mixed deciduous species of low density, poor girth and height. The dry deciduous forests were further classified into several sub-types based on the local classification in vogue. The diagrammatic sketch that follows indicates a broadly acceptable classification of the forests.

---

1 H. G. Champion and S. K. Seth, *A Revised Survey of the Forest Types of India*, Government of India, New Delhi, 1968, pp. 45–47.

2 Ibid.

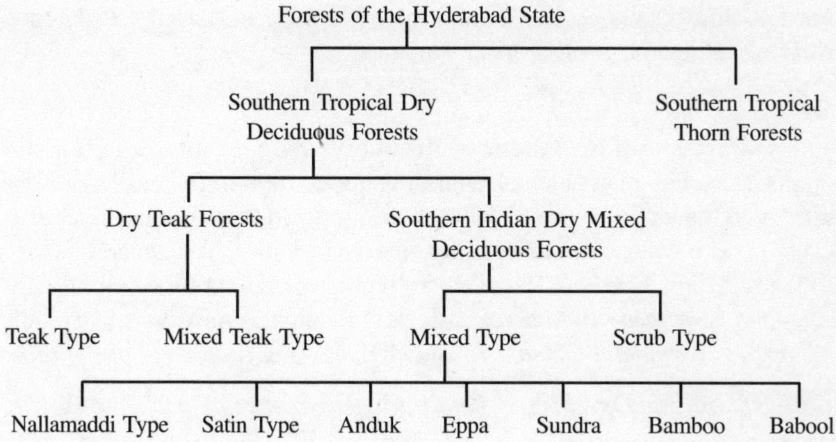

Source: H. G. Champion and S. K. Seth, *A Revised Survey of the Forest Types of India*, Government of India, New Delhi, 1968, pp. 45–47.

Southern Tropical Dry Deciduous Forests

These types of forests were situated mainly in the eastern half of the Hyderabad State. The density of the forest vegetation varied from 0.4 hectare to 0.7 hectare and the average height was 25 feet (7.62 metres) to 55 feet (16.76 metres), with an average rainfall varied from 900–1100 mm. As the growing season is restricted to monsoon (June to September), vegetation generally is not luxuriant and big sized trees of vigour and health were rarely found except in the Godavary and Krishna valleys. These forests consisted mostly of teak (*Tectona grandis*), bijasal (*Pterocarpus marsupium*), tiruman (*Anogeissus latifolia*), nallamaddi (*Terminalia tomentosa*), satin (*Chloroxylon swietenia*), gumpena mohwa (*Madhuca lucida*), sandal wood (*Santalum album*), and eppa (*Hardwickia binata*).

Most of the forest areas were not thick in density and hence various grass species were grown throughout the state. The average height of trees in these forests ranges from 60 feet and sometimes reaches up to 100 feet. The growth was slightly better in the swamp valleys and deep soils, especially along the Godavary basin in the eastern portion of the Hyderabad State where the best quality forest was found. The chief distinction of these forests in the State was that the most important and one of the best teak producing areas of the State fell in this region, particularly in Karimnagar district. The teak well known for its working qualities and durability and defined as the queen of timber, was procured from this region for industrial and defence needs as well as for meeting local demands.[3]

Dry Teak Forests

In these forests teak is found from 10 to over 30 per cent of the floristic composition. The mean annual rainfall is about 1000 mm. Teak is present either predominantly or mixed with other dry deciduous species. Teak also occurs in pure scattered patches in favourable localities. The general quality of the teak is low but the it forms a very high proposition of the crop usually more than 50 percent to 80 percent. Teak is found generally associated with nallamaddi (*Terminalia tomentosa*) and chinangi (*Lagerstroemia parviflora*).

Southern Indian Dry Mixed Deciduous Forests

This group is different from southern dry teak type as here the floristic composition is more represented by species like anduk, tapsi and tiruman. The teak is common in these forests but not as frequent as in the dry teak forests. The forest composition in this type consists of different species with admixture of different levels of growth. There are also intermixed patches of various types of tree species with varying kinds of quality. It is a common feature in the forests of this type that the best quality mixed type or miscellaneous types adjoin a lower quality or even scrub quality forest on degraded and eroded soils. The predominant characteristics in southern dry mixed deciduous forests is that they are generally grown 40–50 feet in height with teak growing in varying proportions up to 10 per cent.

Southern Tropical Thorn Forests

This second category of forests includes the forests consisting of dry mixed deciduous species of low density, poor girth and height. Wherever the thorny species predominate on highly eroded and excessively grazed areas, open type of forests with an average height of 9 to 13 metres can be seen. This type of forests is very common in Western half of the state (in the districts of Aurangabad, Osmanabad, Bidar, Gulbarga, Raichur, Bid and Mahaboobnagar) and is found on eroded ravines, denuded hills and open type of marginal forests. The common species prevalent here are stunted type of *sundra, satin, abnus, bojja, eppa, bor, henkal or danti, menga, tunki, tangedu,* etc.

---

[3] According to Champion, dry teak bearing forests having teak tree from 10 to 30 per cent and pure teak forests over 30 per cent in composition were classified under 5A/G 1 (i.e., sub-groups 5A, Southern Tropical Dry Deciduous Forests and C1 are classified as the Dry Teak Forests), and the Mixed Deciduous Forests having no teak or less than 10 per cent of teak in composition are classified under 5A/C3 (i.e., C3 = Southern Dry Mixed Deciduous Forests).

Distribution of Forests Zones in the State

Based on the topography, rainfall and geological formation, forests of the Hyderabad State can be conveniently divided into three main zones:[4]

Geographical Location of Forest Types in Hyderabad State

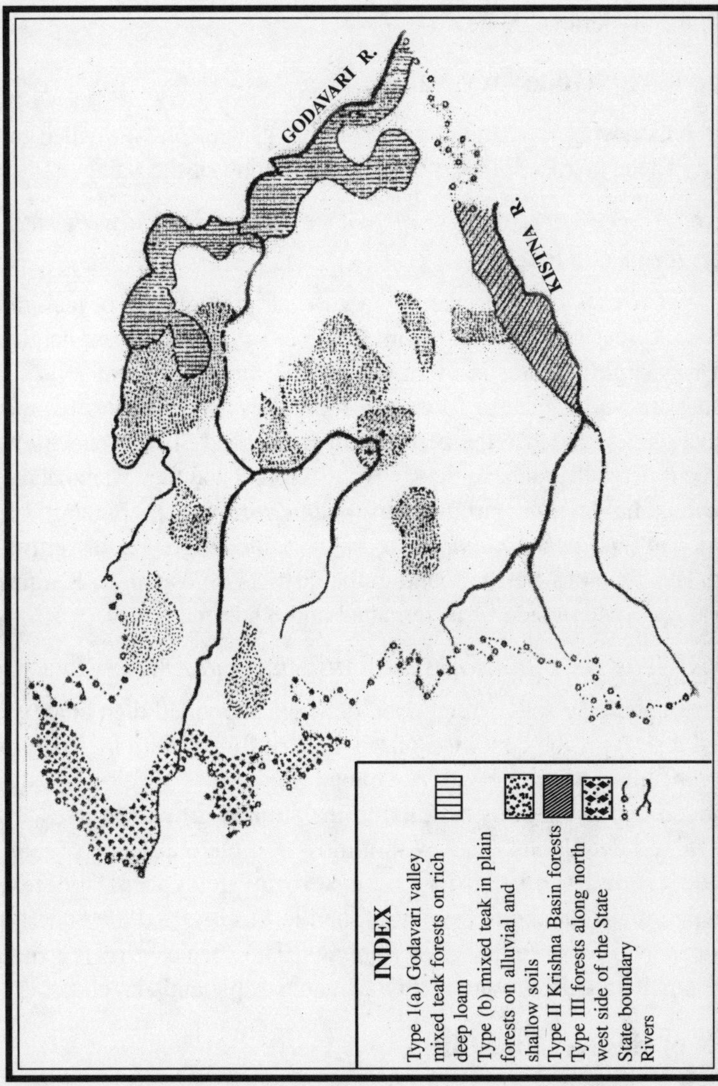

Source: Questionnaire of the Forest Policy Committee, Hyderabad: Forest Policy Committee, 1956, p. 16.

---

4 Government of Hyderabad, *Questionnaire of the Forest Policy Committee, Hyderabad* with forward and explanatory notes issued by B. B. Wadhwani, Secretary, Forest Policy Committee, Hyderabad 1956, pp. 15–16; and Mohammed Sharfuddin Khan, *Forest Flora of Hyderabad State*, Government of Hyderabad, Hyderabad, 1953, pp. vi–viii.

1. forests of the Godavary valley;
2. forests of the Krishna valley
3. along the northwestern side of the State.

The map, given on page 17, shows the geographical location of forest types in the Hyderabad State.

Forests of the Godavary Valley

Along the Godavary river the group of forests grown can be divided into the following two categories, based on the topography of the valley.

1. *Mixed Teak Type Forests (on well drained slopes with rich deep loam soil)*

This type of forests is represented by a good straight growth of teak ranging from 60–100 feet in height and from 4–6 feet in girth at breast-height. The growth was slightly better in the moist valleys and deep loam soils such as the Godavary valley. These forests in the valley were the most important zones that used to produce the best teak in Hyderabad State. Apart from teak, several commercially valuable trees were also found and they were nallamaddi, eppa, abnus, haldava or parpuganam (*Adina cordifolia*), chinangi, bijasal, tuni, inderjan (*Holarrhena antidystenterica*), bamboo, grewia and graveolens species. This type of forest is found in the districts of Warangal, Karimnagar, Adilabad and eastern side of Nizamabad and Khammam.

2. *Mixed Teak (in plain forests on alluvial and dry shallow soil)*

While loam and clay soils with a good drainage supported high quality forest of mixed teak type, the condition of this type of forests with impeded or bad drainage resulted in the growth of stunted mixed teak type of forests. The deteriorated quality of soils resulted in the stunted growth of trees. Poorer quality mixed deciduous types with almost complete absence of teak was found on eroded sandy soils where xerophytic type of species was commonly found. Bamboos were not found in this belt and the undergrowth was represented by different grass varieties. The species grown apart from teak were nallamaddi, tirman, abnus, chinangi, tapsi and sizyphus.

Forests of the Krishna Basin

Along the Krishna river and within its watershed area the eppa types of forests were common. These forests generally consisted of tirman, abnus, nallamaddi and other species. On its northward spread these were merged with type '2' of the Godavary valley forests where on shallow trappean areas eppa became gregarious in patches but never attained large sizes till it merged with type '1' where on sandstone areas it reached its maximum dimensions.

Forests along the Northwestern Side of the State

Here the most common were the sandal (*Santalum album*) type of forests, with their home in the southern peninsular India, forming a belt running north-westwards in the Aurangabad district which finally merged with Khandesh of the Bombay Presidency. Clayey soils with stiff clay content and poor drainage or black cotton type of soils of this belt contained blank patches with grass and were devoid of good quality forest. This type had been very much maltreated and was found scattered. The species that were prevalent in these forests were sandal, tirman, babul, sundra, manga (*Anogeissus latifolia*), amaltas (*Cassia fistula*), narlinga (*Albizzia amara*), satin, neem (azadirachta indica), bel (aegle mazmelos), *Zizyplus jujuba*, tore-elka (*Limonia acidissima*), mori (*Buchanania latifolia*) and kaweet (*Feronia elephantum*). This forest belt gets merged with the tracks in Gulbarga, Osmanabad, Aurangabad and Bhir districts.

The description on the above mentioned forest belts gives a clear indication of their ecological spread in correlation with the edaphic (soil) factors. One forest type merged into another after passing the edaphic margin. At the border of this mixed edaphic margin mixed characteristics of both types of forests could be noticed. The main forests of the Hyderabad State were situated in the districts of Warangal, Karimnagar, Adilabad and Nizamabad. Adilabad district had one of the best teak forests in India.[5] However, there was also considerable inferior forest growth in Medak, Mahaboobnagar, Nalgonda, Aurangabad, Nanded, Parbhani, Gulbarga and Raichur districts. Broadly speaking, the forests of the Hyderabad State may be described as varying from 'good' in the eastern part of the State to practically 'non-existent' in the west. The following table 1 shows the forest area in each district in 1914–15.

---

5 The multi-species nature of these forests requires some explanation. Many important species such as teak, nallamaddi, mohwa, bamboo, etc. were crucial both for rural livelihood and for revenue purposes. There was, however, rarely a direct contradiction between growing trees for timber and the survival of mixed forests. Sometimes the Forest Department intervened in manipulating the forest tracts, but it was, largely not creating monocultures. However, even though if they were successful in creating single-species areas but at the end forests had a mixed growth of different tree species.

Table 1. Statement showing the distribution of forests in the Hyderabad State in 1914–15

Name of the district	Total area (sq. m)	Forest area (sq. m)			% of forest area to total area
		Protected	Open	Total	
Aurangabad	6,212	220.54	63.33	283.87	4.56
Bir	4,132	...	160.00	160.00	3.87
Parbhani	5,127	95.34	63.02	158.36	3.09
Nanded	3,797	76.11	267.66	343.77	9.05
Osmanabad	2,743	...	...	...	...
Gulbarga	6,719	199.41	54.22	253.63	3.77
Raichur	6,791	26.10	142.96	169.06	2.48
Bidar	5,857	...	50.00	50.00	0.85
Total Marathwada and Karnataka	**41,378**	**617.19**	**801.19**	**1,418.69**	**3.42**
Atraf-i-Balda	2,611	N.A.	N.A.	N.A.	...
Medak	3,215	107.49	192.08	299.57	9.31
Mahaboobnagar	5,147	983.10	133.00	1,116.10	21.68
Nalgonda	6,104	1,355.32	109.61	244.93	4.01
Nizamabad	3,273	352.76	146.59	499.35	15.25
Warangal	7,943	2,124.38	926.00	3,050.38	38.40
Adilabad	7,294	1,831.50	982.66	2,814.16	38.58
Karimnagar	733	1,038.68	280.55	1,319.24	23.01
Total Telangana	**38,709**	**6,573.23**	**2,770.49**	**9,343.72**	**26.13**
GRAND TOTAL	**82,698**	**7,190.73**	**3,571.68**	**10,762.41**	**13.43**

Source: Report of the Revenue Administration of HEH the Nizam's Dominions for 1324 Fasli (1914–15), Hyderabad, 1916, p. 128.

Table 1 shows that the forest area under control and management of the Forest Department was estimated to be 10,762.41 sq. miles which was only 13.43 per cent of the total area of the State. In British India the proportion of forest area was 24.5 per cent for the whole of India, 21.8 per cent in the Central Provinces, 12.76 per cent in Madras Presidency and 10.6 per cent in Bombay. The minimum percentage of forest area laid down by authorities on Indian forestry was 20 per cent.[6] When looked from the British standard, the Hyderabad State did not possess sufficient extent of forests. Moreover, the

[6] *Report of the Administration of HEH the Nizam*'s *Dominions for 1324 Fasli (1914–15),* Hyderabad, p. 1916; and B. Ribbentrop, *Forestry in British India,* 1900, Indus Publishing Company, New Delhi, reprint 1989, p. 107.

13.43 per cent of forest area was not at all uniformly distributed all over the districts in the State. It varied from 0.85 per cent in the Bidar district to 3 to 9 per cent in the other Marathwada districts, rising as high as 38.58 per cent in the Adilabad district, whereas one Maratha district, viz, Osmanabad had no forest at all. The Marathwada districts were far less wooded than the Telangana tract; and the 3 Telangana districts of Warangal, Karimnagar and Adilabad possessing more than 70 per cent of the total forest area of the State. Excepting the districts of Nizamabad, Medak and Nalgonda, the rest of the Telangana districts were above the average of 20 per cent, i.e., minimum percentage of forest area laid down by the Forest Department.

Forests in the Pre-1867 Period

Up to the end of eighteenth century, forests were almost regarded as 'open access resources', often for everybody's use. With the exception of sacred groves and forest areas meant for royal hunting, forests were used for all purposes, free to all, although the ownership of the forests generally vested in the rulers.[7] There were indeed certain customary restraints on the use of forest resources, as in the sacred groves protected by villages. Many of these groves survived upto the late nineteenth century, detailed descriptions of which were left behind by the British officials.[8] In the Hyderabad State as well, the practice of venerating trees was in existence particularly in Telangana and Karnataka regions. Some of the eighteenth century regional States in the Deccan followed the policy of claiming property rights over certain species of trees.[9]

---

7 There are different kinds of opinion among the scholars over the question of how pre-colonial situation was different from the colonial period. Gadgil and Guha argue that in the pre-colonial period village communities had control over management and disposal of forests and uncultivated wastelands. Demands by dominant landholders and rulers were limited, and never approached the scale they did in the subsequent period of colonial rule. On the other hand, Richard Grove says that deforestation had assumed significant proportions before the advent of colonialism. He also argues that initiatives for state control of forests in provinces like Sindh anticipated the colonial conservation. See, R. Guha and M. Gadgil, *State Forestry and Social Conflict*, p. 141; and R. Grove, *Green Imperialism.*

8 D. Brandis, *Indian Forestry,* Woking, 1897, p. 12.

9 For example, Hyder Ali and Tipu Sultan of Mysore asserted rights over sandalwood tree and claimed that forest land was the property of the king. See M. V. Nadakarni, *The Political Economy of Forest Use and Management,* Sage publications, New Delhi, 1989, p. 31; for more details, see Francis Buchanan, *A Journey from Madras through the Countries of Mysore, Canara and Malabar,* Higginbothams and Co., Madras, 1870, Vol. II, pp. 135 and 262.

The Nizam rulers showed interest, prior to the establishment of the Forest Department, in following policy of reclaiming the forests and extending the area under cultivation. After the establishment of the Hyderabad State in 1724, their main concern was to consolidate the political power. First, Nizam followed the administrative system of granting *jagirs* to the officials and nobles of the State.[10] He consolidated his power by rewarding *jagirs* to his followers who accompanied him in his campaigns in the Deccan.[11] The dominant political ideology accepted by the Hyderabad State was that 'the State was founded on the principles of pure despotism or autocracy, the whole of being considered as the private property of its Ruler.'[12] It was thus left to the ruler to grant or terminate the grant of a *jagir* at his own absolute discretion. The *jagirdars* were the assignees of land revenue and in addition to it they had also jurisdiction over all natural resources like forests, fisheries, etc. in their *jagirs*.[13]

Even before the establishment of the Forest Department the commercial exploitation of forest produce was not totally absent. Though the locals had free access to forested part as mentioned earlier, the use of valuable goods like timber and its exploitation was under the control of the Revenue Department.[14] Though teak was, to some extent, allowed for local usage, the State retained the control over its exploitation. The State Revenue Department had its authority over timber and disposed off forest produce to local peasants or timber merchants on permits. The following paragraph shows the method of timber exploitation in the State:

> An applicant stated how much and what sort of material he wanted to take out from a nominated area. He had to pay a fixed amount to government went to forest, and cut down all what he wanted under control or otherwise. As regards the honesty of such people, it might be guessed for good or for worse....[15]

---

10 Document No. 2018 and 4337 (kaifiyat-taujih) quoted by M. A. Nayeem, *Mughal Administration of Deccan under Nizam ul Mulk Asaf Jah (1720–48),* Jaico publishers, Bombay, 1985, p. 182. While granting certain areas as jagirs, the natural resources like forests, grazing lands, fisheries etc., were also gone under the control of the jagirdars who held sway over all the natural resources and enjoyed the income derived from them.

11 JDB. Gribble, *A History of the Deccan,* vol. II, Luzac and Company, London, 1896, p. 5.

12 *Jagir Administration,* vol. 1, Office of the Jagir Administrator, Government of Hyderabad, Hyderabad, 1952, p. 3.

13 Ibid. and K. N. Anantha Raman, *The Report of the Land Revenue Reforms Committee, 1958–59,* Government of Andhra Pradesh, Hyderabad, 1959, pp. 55–58.

14 Andhra Pradesh District Gazetteers, *Karimnagar district,* Government of Andhra Pradesh, Hyderabad, 1974, p. 100.

15 *Reply to the Questionnaire,* p. 12.

However, information is not available on how much of the revenue given on timber trade and other forest produce. There was no system of maintaining the records properly by the government prior to the reforms introduced during 1860s by Salarjung I. However, the pressure on teak resources in the State during this period could not have been anything comparable to what was to come with the emergence of commercial interests after the establishment of Forest Department. In addition to this there was no alienation of the local people (peasants/tribals) from the forests in spite of the state ownership of forest during the pre-1867 period.[16]

Systems of Resource Use

The Hyderabad region had been primarily agricultural and the conditions of agricultural production were, to a large extent, linked to the use of natural resources like forests. Being a tropical country, natural resources had been a source to provide crucial inputs into the dominant complex of agriculture and animal husbandry, while also enabling other forms of resource use such as hunting-gathering, shifting cultivation and artisanal production. As elsewhere in India, in the pre-nineteenth century, there were no restrictions on the use of forests in the Hyderabad State. They were traditionally managed and used by the local communities.[17]

However, for early periods only fragmentary references to traditional methods of regulating the use of forests and pasture-lands are available.

---

[16] C. V. L. Narasimha Rao, 'Details of the Forest Department in the Nizam's Dominions', in *Golkonda Patrika,* (Telugu), 9.5.1938. Across the country at all India level also it is evident from the travel accounts and the reports and observations prepared by the British government officers in the nineteenth and the early twentieth centuries. For details see, Francis Buchanan, *Op.cit.,* and E. P. Stebbing, *History of the Forests,* and 'Forest conservancy in India', *Indian Forester,* vol. 19, no. 62, 1893.

[17] For details at regional and at all India level, see Madhav Gadgil and F. Berks, *Traditional Resource Management Systems: Resource Management and Optimisation,* 1991; Madhav Gadgil and P. Iyer, 'On the diversification of common-property resource use by Indian society', in Fikret Berks (eds.), *Common Property Resources: Ecology and Community-based Sustainable Development,* Belhaven Press, London, 1989, p. 248; Madhav Gadgil, S. N. Prasad and R. Ali, 'Forest management and forest policy in India: A critical review, *Social* Action, vol. 33, 1983; Ramachandra Guha, *The Unquiet Woods: Ecological Change and Peasant Resistance in the Himalaya,* Oxford University Press, Delhi, 1989; Ramachandra Guha, 'Pre-history of Indian Environmentalism: Intellectual Tradition', *EPW,* vol. 20, nos. 45–7, November, 1985, pp. 1939–52; Chhatrapati Singh, *Common Property and Common Poverty: India's Forests, Forest Dwellers and the Law,* Oxford University Press, Delhi, 1986; Atlury Murali, 'Whose trees? Forest practices and local communities in Andhra, 1600–1922', in D. Arnold and R. Guha (eds.), *Nature, Culture, Imperialism,* Oxford University Press, Delhi, 1995.

There are some government records remarking on how customary restrictions on the use of forests operated 'over large areas'. Conservation was achieved by customary limitations, on users rather than formal management. As 'the waste and forested parts never attracted the attention of the Rulers', the people had made use of their products. Forest products like timber and medicinal herbs were lightly taxed when exported on commercial lines. The local consumption of these items however was not taken into account.[18] Up to the beginning of the nineteenth century, forests close to villages were enough to satisfy all subsistence needs of the people, and therefore, forests in remote areas from villages were generally not over-exploited.

Since forests and common grazing lands were an important component of the village economy's resource base, traditional boundaries between villages had evolved to regulate their use. Forests within the village boundaries were considered as the common property of the local communities, who had the right to use them. Almost the whole country, both the cultivated and wastelands, exclusive of the largest forests came to be regarded as within the boundary of one or another village.[19] These traditional village boundaries were important part in traditional management, a necessary condition for enabling the inhabitants of a village to evolve a mutually acceptable and sustainable pattern of resource use.[20] The areas within the village boundaries were not small, but large proportions of uncultivable land and wastelands, including hills and vast forests, had been included from ancient times into the boundaries of nearby villages. Prior to the nineteenth century, it may be argued that no regulation of grazing, lopping or other forest resources was needed as the pressure on these resources was considerably less compared to the later period. People generally used to cut branches that were easy to collect and which supplied a large amount of fodder, in order to reduce the number of head loads that they had to carry.[21] Social norms and customs, may have restrained people from over-lopping any tree and from lopping off

---

[18] In a note by M. H. Ali, Working Plan Officer, Western Circle, Forest Department of the Nizam's government, Instalment No. 265, File No. 6/86, Record No. 1095/1338 Fasli, 1927, (Interim Repository, Andhra Pradesh Secretariat, Hyderabad).

[19] Ibid.

[20] *Administration Report of the Revenue Department, 1324 Fasli (1914–15),* pp. 14–15.

[21] Larger branches, therefore, were not cut, neither very small twigs. They also tried to avoid damaging trees in order to ensure a sustained fodder supply. This was achieved by not cutting all of the branches, particularly small ones, of any tree and by selective trees with a 'good crop' of branches on them. Fuel wood was obtained as a by-product of lopping for fodder from deadwood collected in the forests and by lopping bigger branches from parts of the forests.

trees near other people's houses. These norms and customs were effective when there was an absence of formal controls either by the State or by the local communities.[22] Across the State, pattern of forest use and regulation varied with local conditions. One can get an idea of some of the traditional ways of management of forests particularly by tribal population, like the lopping off trees for fodder and collection of food and minor forest products in forest areas from the studies carried out by C. V. F. Haimendorf in 1930s and 1940s in the Hyderabad State. Here the production of tribal economy as well as agricultural economies were directed primarily towards subsistence needs rather than external markets.[23]

Dependence of Tribal and Agrarian Economies on Forests

The indigenous inhabitants of the Hyderabad State can be divided into two main groups – tribals and peasants. Tribal form of land use was shifting cultivation and food gathering. Peasants settled in villages, however, took over parts of the immediate hinterland to develop agriculture along rivers and fringes of forest areas.

Tribal Economy

In the tribal economy each tribal group controlled a certain land in the forest area and its territory as a community property. The tribe as a whole organized the pattern of resource use on their lands, including area set aside for cultivation in a given year. The main communities of tribal population, Bhils and Gonds practised shifting cultivation, as well as hunting and gathering of minor forest products. Wild fruits, roots, tubers, herbs and nutritious flowers and fruit pulp of the *mahua* tree provided much of the diet, making them less dependent on agriculture. Tribal villages, near plains, were also engaged in trades of firewood, animal horns, honey, medicines, and other minor forest products with the non-tribal population. They evolved a number of cultural traditions to ensure forest sustenance, such as long fallows in the cycle of rotation of shifting cultivation, selecting retention of valuable trees such as mango and *mahua* while felling for slash and burn. The belief systems of the

---

22 Warangal Suba, File No. 37 of 1323 Fasli (1940) and Record No. 1354 of 1350 Fasli (1916), Board of Revenue, Hyderabad, Andhra Pradesh.

23 See, CVF. Haimendorf, *The Aboriginal Tribes of Hyderabad,* 3 vols., London, 1943–1948; *Tribal Hyderabad – Four Reports,* HEH the Nizam's government, Hyderabad, 1945; *Progress and Problems of Rehabilitation in Adilabad District,* HEH the Nizam's government, Hyderabad, 1945; *The Gonds of Andhra Pradesh: Tradition and Change in an Indian Tribe,* London/Delhi, 1979 and *Tribes of India: The Struggle for Survival,* Oxford University Press, New Delhi, 1982.

forest communities were strongly grounded in the worship of nature. Religious festivals were aimed to signify both the agricultural cycle and the flowering and fruiting of the forest trees. Furthermore, certain areas were set apart as refugia, as sacred groves and sacred ponds immune to all human interference. Apart from sacred groves, individual species, like banyan, pipal or other totemic plants or animals were also given full protection. All of these practices enabled the forest dwellers to use their resource base in a sustained fashion. The subsistence-oriented, isolationist life of these communities began to change with the emergence of British colonial power in general at all India level from the late eighteenth century and the beginning of the nineteenth century onwards.

Agrarian Economy

Village was the basic unit of administration of the State with its traditional boundaries that converted into fixed boundaries after the establishment of the survey and settlement departments in 1875.[24] Each village had its regular complement of officers who were usually *watandars* or hereditary village officers.[25] They mainly consisted of the *mali patel* (revenue collection) and *police patel* (maintaining law and order); the *patwari (*maintaining accounts of land holdings and land revenue), the *talari (*village peon), a *sethsindhi* (watchman), *neri* or *neredu (*irrigational water supplier), and a *dheir* or a *mahar (*watchman/messenger) in the village establishments.[26] The government as well as the village community paid these village officers. In addition to the remuneration, some of the village staff such as the *neri* and the *sethsindhi* along with artisans such as the potter, the blacksmith, the goldsmith, the carpenter, the barber, the washer-man, etc. received *Balota inams*[27] or lands with only 1/4th of the original assessment.[28] These were given in consideration of the service they rendered to the village community and government officials. These *balotadars* were also given fixed quantity of grain per plough from the peasants at the time of harvest. In addition to these officials, there were *deshmukhs* (village headman) and *deshpanyas* (district accountants), functioned as tax-farmers before 1860, but afterwards they were allowed to

---

24 *Revenue Administration Report,* p. 25.

25 Moulvi Cheragh Ali, *Hyderabad (Deccan) Under Sir Salarjung 1,* vol. 1, Bombay, 1884, p. 124.

26 Ibid.

27 Land grants given for the maintenance of the village artisans and service staffs.

28 HEH the Nizam's government, *Settlement Report of Warangal taluq of Warangal district, 1302 Fasli, (1893),* Government Central Press, Hyderabad 1894, p. 67.

retain their huge landed estates, besides being given an annual pension known as *rusum*.[29]

The basic units of social organization of a village were populations of several endogamous caste groups. The State was predominantly rural and agricultural and the society had been characterized by feudal and hierarchical relations. The caste based village society had developed a number of regulations or customary restrictions to regulate the use of natural resources of the surrounding areas. The traditional distribution of occupations that constituted a number of caste groups, important of these were *karanam, kapu, reddy* and *velama* distributed all over the State, were largely land-owning and cultivating. There was also Muslim population in the State; some of the big *jagirdars* and State officials were Muslims and rest of them were found in almost all professions.[30] The *madigas, malas* and *mahars* were largely labourers of both agriculture and non-agriculture. Some of them also held land as tenants. Tanning and shoemaking were also pursued by many *madigas*. These classes also performed services pertaining to the public needs of the village as a whole. The other professional caste groups were *yadavas*, also called as *dhangars, gavalis* and *gollas*, who were primarily a caste of graziers and cowherds and had a distinctly agricultural background. The *gaundlas* were engaged in the tapping and sale of toddy in the State; the *ausalas* (smiths), also called *vishwabrahmins*, constituted primarily a community of goldsmiths, blacksmiths, coppersmiths and brass-smiths. There were also *vadrangis* (carpenters), *bhois, besta* and *goodla* (traditional fishermen), *chakalas* (washermen), *mangalas* (barbers), *kummaras* (potters), *waddars, bondilis*, etc.

The caste system was operated on the basis of division of society into hierarchical groups. The division of society depending on the labour performed led people to develop specialized skills and insulated them from any competition from outside the community and even outside the village. This in turn led to a more or less sustainable use of natural resources as each community had a sort of monopoly over a particular natural resource. A good example of this kind of system is monopoly of the toddy tapping

---

29 The right of tax collection was auctioned and tax-farmers enjoyed the authority and advantage of collecting from the peasants much over and above the fixed amount that they paid to government. Barry Pavier, 'The Telangana Armed Struggle', in *EPW*, special number, August 1974, vol. IX, p. 1413.

30 *Andhra Pradesh District Gazetteers, Karimnagar District,* Government of Andhra Pradesh, Hyderabad, p. 69.

community (*goundlas* caste) on the tapping of palm trees, even though grown on private lands.[31]

Though there was the existence of different traditional occupations, agriculture was the main occupation of the people, which was oriented towards subsistence needs. The existence of forest resources had provided the resource base for peasants and traditional occupations alike in varied manners. In the village areas, forests provided food, fodder, fertilizer, fuel and timber for construction and implements. Forests were considered as a free gift of nature as perpetual source of village needs. Peasants used manure like farmyard manure, sheep manure and compost, green manure like sunhemp, *daincha, pillipesara* and wild indigo.[32] All these organic manure were either gathered from forest areas or from village grazing lands (*gairans*) and were used in paddy and dry crops.[33] Forest farmyard manure, which was applied prior to the commencement of preparatory cultivation, was of two types. One was lead moulded gathered from forest areas and the other was made by using green and dry leaves which served as litter and bedding. This was mixed with dung and urine and served as manure to the fields. For a couple of months or so manure was moulded from dry leaves and allowed to rot. Thus the forests helped in increasing the nutritive value of the agricultural lands, directly through its foliage and indirectly through the excreta of the cattle fed with fodder, leaves more of paddy grass and forest grass.

Common grazing lands and pastures called *gairans* and un-assessed wastelands (*poramboke* land) were a common feature of villages in the Hyderabad State. These resources were open to all members of the village community. The strength and sustenance of working cattle depended on the type of breed, facilities available for fodder and grazing, etc. The extent of grazing lands that were earmarked for each village fixed at 25 per cent to the total area of a village.[34] There were, however, signified variations following the habitat, topography, size and population of a village as well as number of historical and socio-political factors.[35] Within the village areas, there were

---

31 A typical example of this kind also existed in Maharastra area where butchers were not allowed to stocking sheep. They had to procure them from the shepherds on and when required, see *Imperial Gazetteer of Bombay Presidency, Kolaba and Janjira,* vol. II, 1909, pp. 33–34.

32 *Andhra Pradesh District Gazetteers,* Karimnagar District, p. 89.

33 Ibid.

34 A. I. Qureshi, *The Economic Development of Hyderabad,* vol. 1, Orient Longman, Madras, 1941, p. 78.

35 Due to the schemes of the expansion of cultivation, these pasture lands came to be allotted for agricultural development and got shrunk largely from 25 per cent to 5 per

other types of common lands also. These were classified as barren and uncultivable lands, cultivable wasteland, land for non-agricultural use and so on. Except the categories where the ownership (*patta* lands) was otherwise defined, all the common lands were owned by the government and administered by the Revenue Department.[36] The local people usually had rights over them.[37] In Telangana area, their extent was about 25 per cent of total village area, whereas in Maharastra region, the extent was greatly degraded to 5 to 20 per cent of a village land due to low forest cover.[38] In addition to cattle, professional graziers like D*hangars, Gollas* and *Gavalis* also fed sheep and goats on pastures and forests. The herbage on the forest floor provided grazing and the forest floor was burnt by them (peasants and shepherds) in the spring of every year to remove the heavy litter that tends to choke grass growth of grass on which livestock grazed.[39] So forests and pastures were an important source of food for livestock. Thus, the agricultural economy along with the animal husbandry was crucially dependent on forests and common lands. The wood, which was not fruitfully used for construction work or making implements or articles, was largely consumed as firewood.[40] Bamboo trees were used for building purposes and making trays of various types apart from household articles of daily use.

---

cent of total area of a village. Even these much reduced areas were also sacrificed whenever lands fell short in making land grants on patta to maqtedars, ijaradars and kowldars, see *Report of the Agrarian Reforms Committee,* Government of Hyderabad, Hyderabad, 1949, pp. 88–92. Consequently in later periods, the peasants were forced to raise fodder crops among which *kadbi* or jawar stalks, was the most important, *Imperial Gazetteer of India, Hyderabad State,* p. 32.

36 M. H. Ali's note, op. cit.

37 The concept of ownership in use seems to be very widespread among preliterate people, representing the sense of ownership which merges, with the full support of custom, as a result of constant use by a particular individual or family; *Anthropological Glossary,* Robert E. Krieger Publishing Company, Molabar, Florida, 1985, p. 266.

38 M. H. Ali's note, op. cit. Marathwada, being a dry region pasture lands were subjected to the vagaries of monsoon. Hence, neither the quantity nor the quality of fodder was significant enough. The principal means for grazing were to private lands which were available free after the first crop was over. After the harvest, the cattle grazed freely on the crop stubble.

39 M. H. Ali's note, op. cit. This system prevailed in most of the forest areas of India. In the part of the Himalayan region on this issue, see Ramachandra Guha, *The Unquiet woods,* pp. 28–29.

40 Most cooking fuel in the villages of the Hyderabad State was firewood. Firewood gathered by villagers was mostly deadwood or branches lopped from trees. Felling for this purpose was rare and usually illicit. But trees were felled for timber for houses, cowsheds and agricultural implements.

Gum, wax, honey and lac were also exploited. The forests also supplied spices, medicinal herbs, tamarind and leaves of certain species of trees that were used for making *beedis*, leaf plates, baskets and rain covers.[41] Thus in various ways agrarian life in the State was closely tied up with the forests. Villagers' awareness of the importance of their forests and their concern for them was reflected in the region's folk songs.[42]

There were not many records on how customary restrictions on the use of forests operated 'over large areas' in the Hyderabad State. Conservation was achieved by customary limitations on users rather than formal forest management. Peasants' dependence on natural resources was institutionalized through a number of social and cultural mechanisms. Prior to the establishment of the Forest Department, forest and animal care, welfare and protection were part of the religious rituals. They were more the traditions rather than enactments. Through religion, customs and folklore the local communities had conserved natural resources in many ways.[43] Interestingly, certain traditions actually aided conservation of natural resources like the belief that certain trees and plants like banyan, pipal and tulsi, were treated as sacred and thus conserved. *Devarayas* or temple groves were venerated and protected from injury, which were spread from a few trees to forest area of several hundred acres. Deitrich Brandis, the first Inspector General of Forests in India, said, such sacred groves were found all over the country, and no villager would cut the vegetation in any way.[44] In the case of the Hyderabad State also there were many sacred places in forest areas. Many hilltops were the abodes of local deities and the vegetation around the hills was regarded with great respect.[45]

---

41 *Andhra Pradesh District Gazetteers,* Karimnagar District, pp. 99–100.

42 J. Thirumala Rao, *Telangana Raitanga Poratam: Praja Sahityam* (Telugu), Sahiti Circle, Hyderabad, 1988, reprint 1995. For more songs refer to, P. Sivaramakrishna, *Telugu Girijana Geetalu* (Telugu), unpublished PhD thesis, Osmania University, Hyderabad, 1982.

43 Ramachandra Guha and Gadgil have strongly supported these arguments in their studies. The protection of cattle was important in shaping the practices of mixed agriculture and animal husbandry, so characteristic of India. Traditions such as these, according to Guha, have played an important role in influencing the pattern of utilization of biological resources in India over many centuries especially until the arrival of Europeans; see, *The Unquiet woods; Forestry in British and post-British India*; Gadgil and Guha, 'Greening the commons'.

44 D. Brandis, *Indian Forestry,* Woking, 1897, p. 12.

45 For example, the hilltop of Yadagirigutta in the Nalgonda District is the abode of Lakshmi Narasimha Swamy and the surroundings of this hill is lushed with thick forests; in Khammam district also, in a place near Papikonda Hill, a rain God and a

Another tribe Chenchus had traditional customs and ritual practices, such as the first fruit offerings to *Garelamaisamma*, the deity of the forests and the chase in the Nallamala hills on the either side of the river Krishna. The clan deities of Gonds, the *Persa Pen*, were thought to have acted on the protection of the members of each clan; thus they were venerated with great respect.[46] The *Keslaput jatra* (religious fair), the most popular religious event in Adilabad district was performed in honour of the clan deity of the *Buigoita* branch of the *mesram* clan. Another ritual focal point of the entire celebration was a simple sati shrine, erected on a mould close to a huge banyan tree.[47]

Thus, people had developed a variety of social arrangements in the Hyderabad State suitable to the prevailing conditions to ensure that they did not suffer from a scarcity of the forest resources on which they were so dependent. 'The non-cultivated tracts of the country were firmly committed to a role of supportive of agriculture and were maintained well by the villagers as assets under community control.' Several researches at all India level also reveal that the non-cultivated tracts in the villages were used for supportive role to agricultural activities and were maintained as village assets under community control.[48] The absence of any reports of shortages of forest produce in the period preceding the establishment of the Forest Department and the advent of commercial forestry suggests that these methods were generally successful in resolving the problems of the natural resources.

---

female deity were venerated with great respect; *Bhimana* shrine in forest area of Mahur in Adilabad district was venerated by tribal people of Kolams, Gonds and Naikpod clans. The belief of Hill Reddis in Khammam district on the *Konda devatha* for the prosperity of the crops and for the successful cutting of bamboos can be considered as traditional way of exploiting forest resources. CVF. Haimendorf, *Tribal Hyderabad,* pp. 25–26 and 187–188.

46 Ibid., p. 169.

47 CVF. Haimendorf, *Tribes of India,* pp. 167–184.

48 see, Madhav Gadgil, 'Forest management, deforestation and people's impoverishment', *Social Action,* vol. 39, Oct–Dec, 1989, pp. 357–383.

CHAPTER 3

Evolution of Forest Policy

For a long time, prior to the colonial period, forests had no commercial value to the rulers and the forest produce was predominantly consumed by the forest-dwellers and the peasants. Traditionally forests were conceived as an unproductive resource and the people living nearby the forests were allowed by the states to exercise their traditional customary rights. The Hyderabad State had also not shown any interest to conserve the forest resources till the middle of the nineteenth century. No forest policy was initiated when the Asaf Jahi rulers commenced to govern the Hyderabad State, nor it was realized as elsewhere in India that such a policy was considered necessary. With the exception of a few trees used for commercial purposes such as teak, the rural communities had free access to the forests. Forests were seen largely as an obstacle to the spread of cultivation rather than a valuable resource. Like their Mughals predecessors, the Asaf Jahi rulers also tried to push back the forest areas and extended the area under cultivation. The success of land revenue policy of the State was often measured in terms of extending the agrarian frontier.

As a result of this agricultural expansion during the first half of the nineteenth century, the forest areas of Warangal, Adilabad and Karimnagar districts situated within the Godavari basin, had been repeatedly scratched up and excised under *ijaras, banjars* and *pattas*.[1] As a result best deep and

1 Nizam's government, Revenue Department, File no. 131/86, Record no. 420/1351 Fasli, Instalment no. 219, 221, 1941, Interim Repository, AP Secretariat, Hyderabad. The *Ijara* tenure was introduced during 1850s during the premiership of Salarjung I. It was found that owing to the maladministration, land had considerably diminished in value, large

alluvial soil that would normally grow big timber was first cleared for cultivation. The vast range of deforestation led to big environmental problems such as soil erosion by the beginning of the twentieth century.[2]

Political/Administrative Causes

In the middle of the nineteenth century, the British after consolidating their hold over India began to lay down an extensive railway network. Great chunks of forests were cut to meet the demand for railway sleepers. No supervision was done over the felling operations and a large number of trees were felled whose logs could not be utilized. The devastation that was wrought by private contractors, finally awoke the British administration to the fact that India's forests were not inexhaustible.[3] In 1862, the viceroy called for the establishment of a department that could ensure the sustained availability of the enormous requirements of the railways for sleepers. It was in this situation that the Imperial Forest Department was established in 1864.[4]

As a result of the establishment of the Imperial Forest Department, the large areas of forests under the princely States were also drawn in, though indirectly, into the orbit of colonial commercial exploitation. The exploitation of these forests, either by the Imperial Forest Department, or in the later years, directly by the native rulers was instrumental in introducing the commercialization of forests. Over the time, like other native rulers, the Nizams, who were indifferent towards their large forest wealth, became very much sensitive to the value of their forest properties.[5] With the growth of neighbouring markets in British India and with the construction of railways in between 1850s and 1860s huge chunks of forests were needed to meet the demands for railway sleepers and fuel. Thus the importance of forests was increasingly felt by the Nizam's government.

---

tracts of culturable land were laying waste, and numerous villages had been deserted. Special inducements were taken to attract people to the land, so as to re-populate the deserted villages and bring the forests and wastes under cultivation. To achieve this object the *Ijara* system of leasing out lands was introduced. *Revenue Administration Report, 1914–1915,* p. 53.

2 *Report of the Administration of HEH the Nizam's Dominions, 1331 Fasli,* (Companion Volume), 1920, p. 21.

3 Ramachandra Guha, *Forestry in British and Post-British India: A Historical Analysis*, Part I, *EPW*, vol. 28, no. 44, 29 October, 1983, pp. 1882–1896.

4 Ibid.

5 Ibid.

In 1860, Afzaluddaula, the Nizam V, alienated the land required by the Bombay-Madras Railway line passing through the Hyderabad State in the Raichur Doab. The Nizam-State Railway was established fourteen years later in 1874 and progressed gradually in a phased manner and at the time of their merger with the Indian railways, the Nizam State Railway had a length of 1938 miles.[6] Though the Nizam State Railway was started in 1874, the demand for teak was generated much before to this period primarily due to the demands from the railways of British India.[7] In other words the commercial exploitation of the forests in the Hyderabad State predates the formation of forest policy and forest department. The Godavari valley, particularly, harboured the finest stands of teak found anywhere, with the tree's fantastic wood being in great demand for railway sleepers. During this period timber merchants and local contractors appeared on the scene. Their promiscuous greed devastated the large forest areas in the Godavari valley. For instance, timber merchants and contractors, by taking permits from the revenue department, carried out reckless exploitation of the forests of the Karimnagar, Adilabad and Warangal districts during the 1850s and 1860s i.e., before the advent of strict government control. A far greater number of trees were felled than actually required and the excess wood felled was exported by floating down in the Godavari river into the areas of the Madras Presidency.[8] But with the improvement in road network in later years more parts of the forest area were opened up to the needs of the larger economy.[9]

---

6 B. K. Narayan, *Finances and Fiscal of Hyderabad State (1900–1956)*, Government of Andhra Pradesh, Hyderabad, 1973, p. 62.

7 The expansion of railway network in India generated a high level of demand for timber. The railways required more than one million sleepers every year in the initial stages and each sleeper lasting normally 12 to 14 years. A mile of broad-gauge railway track required about 1800 to 2000 sleepers. On an average one mature tree (teak, sal and deodar, etc., being the only species of appropriate strength) could provide 7 sleepers. Until 1890s, when the use of coal began to become general, the forests also had to provide for railway fuel. See, D. Brandis, 'Memorandum on the supply of railway sleepers of the Himalayan pines impregnated in India', *Indian Forester*, vol. IV, 1879, pp. 365–383, especially p. 366; also quoted by Ramachandra Guha, 'Forestry in British and post-British India: A Historical Analysis', pp. 1882–96.

8 Nizam's government, Political Department, File No. 6/86, Record No. 1093/1338 Fasli (28.10.1927), AP State Archives (APSA); and *The Report of the Administration of HEH the Nizam's Dominions, for 1331 Fasli*, Companion Volume, 1922, Hyderabad, p. 32.

9 The building of the large railway network system of the Indian Empire was a major windfall for the Indian timber business. The story of railway deforestation in India is well known, for details see, Ramachandra Guha and Madhav Gadgil, *This Fissured Land*, and R. P. Tucker and J. F. Richards (eds.), *Global Deforestation and the 19th century World Economy*, Duke University Press, Durham, N. C., 1983.

The state of destruction caused by private contractors was admitted by the Nizam administration:

> The net result which is obvious is forests where high quality timber was well known to exist are completely depleted of not only big sized and high quality timbers but for want of enough growing stock the very soil is degenerating.[10]

Realizing this alarming condition, the government felt that at least valuable species like teak need to be saved from over-exploitation and this was sought to be achieved by placing conservation of the above species under a separate government agency. Such alarming condition of the forests at another level also legitimated the extension of government control over timber extraction and trade, and also privileged one specific end use for wood (for commercial purposes) over rural needs. Besides this state administration had taken some measures to protect some valuable species for fuel and timber and for their regeneration and development. At this time at all India level the discussions were being held over the environmental effects/benefits of forests. There were also similar discussions in the Hyderabad State for imposing forest regulations and restrictions. The growth of forest exploitation on commercial lines gave rise to various obstacles and problems and the acute need for regulations and restrictions.[11] It gradually became evident that the State could not just allow anybody to extract wood as they wanted. It was felt necessary to enact rules which would organize and regulate timber extraction by permit holders and others. However, there was rarely any investigation into people's rights that were in practice prior to this period. Gradually it was clearly assumed that local populations had never followed any rules and that they were by nature destructive and had therefore to be stopped. At this stage it was important for the state to prepare written rules which would limit the rights of others to the resources and to develop a strong body of foresters to enforce those rules. In a view to develop the Hyderabad State on the lines of the British, Salarjung I, taking all these

---

10 HEH the Nizam's government, Political Department, File No. 6/86, Record No. 1093/ 1338 Fasli (28.10.1927).

11 Local communities were also blamed for destroying teak trees to obtain leaves and timber for use in the construction of houses. In parts of Karimnagar and Warangal districts forests were reduced to 'mere scrub' to meet the rural demand for timber and fuel. Peasants were suspected of destroying timber, a productive asset, and of degrading the forest by disturbing regrowth of timber trees. *Replies to the Questionnaire* to Professor E. P. Stebbing's *History of the Forests in India,* by the Nizam's government, 28.10.1927, Inst. No. 265, File No. 6/86, Rec. No. 1093/38F, APSA, Hyderabad.

aspects into considerations, established the Forest Department on 16 July 1867.[12]

British Influence

The Forest Department of the Hyderabad State was established on the British model and obviously at that time it was influenced and manned largely by the British. It is significant to note that the three important and key branches i.e., revenue, finance and police were kept under the control of British civil servants which naturally pointed to a strong British control over the State administration.[13] The Hyderabad State was inspired by the neighbouring states that also established Forest Departments. In fact, several princely States like Tehri Garhwal, Tripura, Mysore and Travancore also started their own Forest Departments under the broad policy framework of British forest management. Since the Hyderabad State was not acquainted with the principles of state forestry, the Forest Department was kept under the control of British officials, who in turn were helped by the German expertise. The service of the seven British officers as the conservators of forests in the initial years of its establishment shows the presence of strong colonial interference/domination in the state forestry.[14]

Initial Administration

The first task of the Forest Department was aimed at checking the reckless deforestation and to identify the sources of supply of strong and durable timbers such as teak and nallamaddi (*Terminalia tomentosa*). Protection of forests began as a response to the depletion of large sized teak trees in the forested districts of Telangana. Initially it was largely a question of enforcing a minimum size to be cut and rules were framed for regulating the lopping of trees for railway sleepers and fuel, while trade by people in any form of forest product was prohibited.[15] People could exercise their rights subject to

---

[12] Nizam's government, Revenue Department, A Forest Note sent by The Revenue Department to the Secretary of the Constitutional Affairs, 1941, File No. 131/86, Record No. 420/1350 Fasli, Instalment No. 219 and 220, (Interim Repository, AP Secretariat, Hyderabad); also in, *The Statistical Year Book of HEH the Nizam's Dominions for 1349 Fasli (1940),* Hyderabad, 1943, p. 153.

[13] M. Narasinga Rao, *50 Years of Hyderabad,* Hyderabad, 1977, p. 90. It was so pervasive and complete that even matters concerning the royal family such as the wedding of the princes, their education and succession to the throne, the opinion and advice of the British Resident carried all weight and approval.

[14] *The Statistical Year Book, HEH the Nizam's Government, 1351F to 1354F (1941–42 to 1944–45),* Secunderabad, 1949, p. 646

[15] *Report on the Administration of HEH the Nizam's Dominions for 1304 to 1307 Fasli (1894–98),* vol. 1, Report, Madras, 1899, pp. 71–73.

the rules prescribed in these elementary restrictions. In practice the protection to timber trees was more crucial than the half measures for establishing government control of timber-bearing tracts of land. The effectiveness of protection for valuable timber was limited because only a small establishment in its initial stages could do very little to control extensive forest tracts.

In 1881, the Forest Department was transferred from the home secretariat to the Revenue Department. Between 1886 and 1893, the Forest Department under the control of Mr J. Ballantine as the Conservator of Forests, made rapid strides in the demarcation and improvement of forests. The government felt that for effective functioning of the Forest Department, the previously exercised rights of the people should be restricted. All this was done in the name of conservation of forests and the reservation of forests was of course done in the name of public benefit. The foundation stone for the state forest policy was laid in 1890 envisaging preservation and improvement of the existing forests through the regulation of customary rights and imposing restrictions on the traditional privileges of the people. Some other features of this policy were like this:[16]

1. Forests in some areas should be preserved to maintain climatic and physical conditions.
2. Good quality forests should be exploited on commercial basis as a source of income to the State and
3. Giving away of forest lands to the development of agriculture should be properly checked and should be relinquished with certain conditions.

In response to the strict adherence of forest policy, the government in 1894 notified 3390 sq.miles of forest area in Warangal, Karimnagar and Mahaboobnagar districts as the protected forests and placed them under the sole charge of the Forest Department. The rest remained un-notified. This notification marked the commencement of a new era in the forest administration of the State. Some important rules known as *The Reserved Stateforest Rules* came into being for administration of these areas from 1894 onwards. These rules consisted of two parts: the first relating to those forests that were demarcated and the second to those that were not yet demarcated. Some of the important features of these rules which was reflected in the state forest policy, were that the management of the demarked reserves came directly under the control of the Forest Department with full powers to frame and

---

[16] Nizam's government, Revenue Department, Circular (record) No. 420/1350 Fasli, File No. 131/86, Instalment No. 224, 226, 228 and 230, 1941, (Interim Repository, AP Secretariat) Hyderabad.

issue regulations on preservation and exploitation. As per the regulations, people were supposed to obtain prior permission to fell or remove trees in demarked forests, if failed they were liable to be punished. Unrestricted grazing was prohibited and in certain areas grazing was permitted with a fee (the grazing fees were fixed at two annas per buffalo, one anna and six paise per ox or cow and one anna each per goat or sheep).[17] Restrictions and limitations on cultivation were imposed on dry and wet cultivation practised in demarked forest areas depending upon the years of cultivation. People living inside the reserves could use *gyri* (non-reserved timber), grass and other inferior kinds of forest produce for domestic purposes. But for those living outside the reserves, *gyri* timber was given on payment.[18]

In addition to these rules there were also some other restrictions imposed on the common people such as hunting, removing of minerals, kindling fire, etc., inside a reserved forest. These activities could not be carried out without the permission from the Forest Department. In the case of un-demarcated forests, the grazing fee fixed at half of the demarcated forest areas. With regard to cultivation and forest produce it was laid down that until the forest had been declared a reserved forest, 'no restriction shall be placed on the extension of cultivation around villages, except where there is a valuable timber, and the ryots shall continue to possess those rights in connection with grass, timber, grazing, etc., which they at present enjoy in forests nor reserved; with this difference only, that they shall be controlled by the Forest Department, instead of by the Revenue officers.'[19] However, these restrictions were not so stringent compared to the demarcated reserves.

As a part of reforms in the Hyderabad State, the Legislative Council was established in 1894 for the purpose of enacting laws for the state.[20] During the same period the Hyderabad State government was arming itself with

---

[17] The revenue from grazing in the reserves had increased by leaps and bounds since the rules for the reserves were issued in 1895. In that year the amount realised from grazing was Rs. 1,412 in the next year when the rules were fully applied the revenue rose to Rs. 15, 804. In the famine year of 1897, notwithstanding concessions granted to peasants, the collections rose to Rs. 35,065 and in 1898 they stood at Rs. 44,107. These figures show the commercial outlook of the Forest department; see, HEH the Nizam's government, *Administration Report of the Forest Department* (ARFD), *1894–95,* p. 74.

[18] HEH the Nizam's government, Revenue Department, *The Reserved State Forests Rules, 1303 Fasli,* (1894), Transfer List, Instalment No. 38, File No. 64, G1/ C–21, APSA, Hyderabad.

[19] *ARFD,* 1893–94, pp. 70–71.

[20] M. Fathaulla Khan, *A History of Administrative Reforms in Hyderabad State,* Secunderabad, 1955, p. 116.

necessary legislative powers by strengthening the Forest Department and the restrictions imposed on forest use acquired the sanction of legislation. In the *Annual Administrative Report of the Forest Department of 1899–1900*, the Conservator of Forests said that the Draft Forest Law was submitted to government in 1893 as it had become impossible for him to enforce the rules of the Forest Department, aided only by the circulars of government. These circulars did not provide the punishment for offenders who were quite aware of the fact that they could evade or violate departmental rules with impunity.[21] The Forest Bill, drawn on the lines of the Berar Forest Law,[22] was introduced into the Legislative Council in 1897 for enactment. Commenting on this bill, *The Deccan Post*, a weekly newspaper largely dealing with the Hyderabad affairs and published from Bombay, said:

> It cannot be denied that the Conservator has just cause of complaint under this head, for in the absence of any provision for punishing offenders against Forest Regulations. The forests of the State cannot be secured against exploitation and havoc, but at the same time it must be admitted that the draft before it passes into law required thorough sitting and all arguments *pros* and *cons* should be exhaustively discussed. Government always have two objects prominently before them: one is the comfort of their subjects and the other is protection of their forest from reckless waste. As the measure is the first of its kind no amount of its labour and protection now paid to it can be said to have been wasted, and there is every hope that the two-fold object of government will be fully attained by having a forest law for HEH the Nizam's Dominions which will not be a source of worry and vexation to any one upon whom it bears.[23]

The draft had met with much opposition particularly upon the provision of reservation of lands which would alienate people from having access to the forest areas as was done in the British India. However, the domination of the state appointed members of the Legislative Council silenced the criticism

---

21 ARFD, 1898–99, p. 7.

22 Several provinces drafted numerous sets of local rules and brought into force after the Indian Forest Act of 1878. These varied in different parts of the country and had been added to or modified as and when necessary. According to the Berar Forest Law (1886), all wasteland in Berar which had not been alienated under settlement was the indisputable and entire property of the State. Hence the creation of state forests became purely a selection and the sections treating of this creation were very simple. See, B. Ribbentrop, *Forestry in British India,* 1900, Indus Publishing Company (Reprint), New Delhi, 1989, p. 102.

23 *The Deccan Post,* 15.6.1900, p. 6.

and almost all the provisions of the bill were approved. As told earlier that the existence of the Legislative Council was only nominal and the general public did not have any voice in it. Upon the introduction of the forest bill in the Legislative Council, *The Evening Mail*, a tri-weekly, published from Bangalore, commented that it would not sit silent in regard to a fundamental clause, which entitled the minister or Forest Department to declare any land as the reserved forest. It protested emphatically against such a drastic measure and hoped that every independent member would consider this point carefully. It also stated that: 'The idea of any one subject of His Highness being able to alienate the land of any other subject possessing equal rights with the other, is too sublime indeed for merriment. We hope that the course suggested for legislation will be opposed strenuously. The *modus operaundi* of the British, when a public exigency arises to adopt a similar course, is to pay the tenant or landlord in full for the hand taken out of his possession or an equal area of land given elsewhere in lieu thereof. Unless a corresponding provision is tackled on to the new law proposed for the delectation of the peasantry belonging to these Dominions. The injurious clause in question ought not to be permitted a niche in the statute book of this State.'[24]

Commenting on the condition of the Hyderabad State forestry, B. Ribbentrop, the IGF to the Government of India, said in his *Annual Report (1899–1900)* that 'The introduction of a suitable forest law or regulation, under which forests can be settled and protected, is apparently an urgent need in this State, and until this is done and the department properly organised and equipped forest administration in Hyderabad will remain unsatisfactory.'[25] Suggesting for strict control over forest areas, W. G. Gilmore, the Assistant Conservator of Forests, Warangal division, said that "Many of the people of Mahadeopur and Kanigeri are beginning to understand the objects of fire conservancy and rendered willing aid; still, without powers to punish incendiarism, trespassers, etc., successful protection is very difficult indeed."[26] Commenting on the regulation of forest areas, E. A. Patridge, the Assistant Conservator of Forests of Aurangabad division, advocated strict supervision and regulation of grazing.[27] In addition to it, the

---

24 *The Evening Mail,* 11.11.1899., p. 3.

25 *ARFD,* 1907–1908, p. 7.

26 Ibid., p. 16.

27 In his words, where he reported from Aurangabad, 'The forests are deteriorating by such frequent and uncontrolled felling, and it is one of the urgent needs in forest conservancy to put a check on such wasteful extravagance; such a reform is needed as will regulate and restrict these concessions to ryots, and the most effective measure would be to place a small money value on such produce, where free grants are now made.', in Ibid., p. 8.

Forest department always protested over the dual control of forest areas (particularly in open forests) and over the lack of coordination between the Revenue and the Forest Departments. The Conservator of Forests said that, "It is only in Reserves that all forest produce is under the control of Forest department. In open forests, all minor forest produce except bamboos is auctioned by Revenue department, generally with unsatisfactory results as they neither tour in the forests nor are they acquainted with the uses and value of the produce sold."[28] During this period large irrigation schemes were initiated by the Revenue Department in forest clad areas of the state in Adilabad district. In this area large scale deforestation for cultivation and colonization was under contemplation in spite of the protests by the Forest Department. In addition to this, it is obvious that when the Forest Department was constituted in 1868, condition of easily accessible forests was extremely poor and chaotic. The ideas for forest management and administration were also in vogue. The drafting of forest laws for the Hyderabad State was contemplated in the last two decades of the nineteenth century, as the regulations were found insufficient. The first important aspect which came up for settlement was, to what extent the long continued right of user to the free collection of minor forest produce such as fuel wood, bamboos, grass, grazing and shifting cultivation in the forest tracts, should be regarded as constituting a prescriptive right. On the other hand, it was acknowledged that government as the guardian of all public interests must insist on the regulation of these rights so as to initiate a good management of the state forests in the Hyderabad State. However, the tinge of profit motive prevailed in these measures. Against this background the first act relating to the forests of the Hyderabad State, the Forest Act II of 1309 Fasli (1900) was passed.[29]

Hyderabad State took Berar Forest Law, which in turn was based on the Indian Forest Act of 1878, as a model for enacting the Act No. II of 1900.[30] The following sentence reveals the fact that, 'Forest legislation in His Highness Dominions follows practically same lines as legislation in British

---

28 ARFD, 1900–01, p. 19.

29 *Questionnaire of the Forest Policy Committee,* Hyderabad with foreward and explanatory notes issued by B. B. Wadhwani, Secretary, Forest Policy Committee, Hyderabad 1956, Government of Hyderabad, p. 20.

30 Mr Ballantine, the Conservator of Forests (1887–1893), drafted Forest Act II of 1900 on the lines of the Berar Forest Law, which was based on the British Forest Act of 1878. In 1886 the Berar Forest Law was sanctioned, according to which all wastelands in Berar were declared as the indisputable and entire property of the State; see, *ARFD,* 1893–1894, p. 71.

India.'[31] Hyderabad State Forest Act was passed in consonance with the Berar Forest Law and which contained all the major provisions of Indian Forest Act of 1878 except minor exceptions.

Forest Act No. II of 1309 Fasli (1900)

This Act empowered the government to declare any land covered with trees, brushwood or jungle as the government forest. Though there was a tradition of state ownership of forests, the Act formalised it by giving a juridical force. The government was empowered to make rules relating to the preservation of trees, forests etc., and also empowered to prescribe punishments for the breach of the provisions of the Act. Elaborate rules were laid down for confiscation of implements used in infringing rules and arrest of offenders.

However, the process of reserving certain trees and areas, for the exclusive use of the government had started much earlier.[32] Though the first Forest Act came into existence in 1900, the process of conservation of forest vegetation had begun even much before in the Hyderabad State. The first attempt had been made in a way to reserve 16 species of trees. The government had authorized the Forest Department to assume the right of reserving the important species and the right of felling timber. The Forest Act No. II of 1900 formalized this process and classified forests as Reserved, Protected and Open or Unclassed forests, based on the forest legislation in British India. This classification was based on the amount of restrictions imposed and the degree of supervision exercised by the government.

Though the Forest Act No. II of 1900 in Hyderabad State came into operation immediately after its enactment, it was not implemented on a proper foothold. Under this Act, no forests had been constituted as reserved forests in accordance with the provisions lay down therein. The *Revenue Administration Report* commented in 1915 said that, 'The Act remained a dead letter and although it has been existed for the last 15 years, no forests have yet been constituted *Reserved forests*. The revision of this Act is at present under the consideration of the Legislative Council.'[33] Poor establish-

31 *Revenue Administration Report,* p. 130.

32 Prior to the Forest Act No. II of 1900, it was intended to form only one class of demarcated reserve state forests and provide a limited protection to all other government forest lands until the time arrived in any particular area to decide which particular forest area should be demarcated and constituted as reserve forest. A reserve forest once notified was safeguarded from infringement of private rights and privileges. HEH the Nizam's government, *Reserved State Forest Rules,* Instalment No. 38, List 2, File No. 64, APSA, Hyderabad.

33 HEH the Nizam's government, *Revenue Administration Report,* p. 130.

ment, insufficient staff and lack of coordination between the Revenue and the Forest Departments were the reasons behind this condition which resulted in poor surveillance and protection of forests.[34] Though the constitution of reserved forests was sanctioned as early as 1894, the survey and settlement of forest areas, which was vital for reservation of forests, started only in 1914 by Claudius of the Survey of India, Dehradun.[35] The only progress made towards the creation of reserved forests was the preparation of maps and boundary description for forests to be reserved till 1914.[36] Although there were no reserved forests technically, several areas were marked as protected forests, over which a strict control was being exercised. They were being managed with the ultimate object of being converted into reserved forests.[37] The forests which were declared reserved before the passing of the Forest Act II of 1900, were also included in the category of protected areas pending the completion of the preliminaries laid down by the Forest Act. The area of protected forests and open forests in 1915 were 7190.73 and 3571.68 sq. miles respectively making a total area of 10,762.41 sq. miles or 13.48 percentage of forest area to the total area of the state.[38]

With the efforts of F. A. Lodge, the IGF (1914–1921), the actual forest areas in the state had been demarcated, excluding as far as possible all occupied villages and their occupied lands. The demarcated forests were placed entirely under the Forest Department's control.[39] In this process, Forest Act II of 1900 was found inadequate and was also revised by revenue officers with the result that it was thought quite unworkable to the standards of the Forest Department, and although it had been in force for 15 years not a single forest had been constituted as 'reserved forest' under it. In the year 1914–1915 the Forest Department was reorganised with an Inspector General of Forests and two conservators under him. In the same year the Forest Act II of 1900 was amended and made more effective, laying down detailed procedure for reservation and for controlling the incidence of forest offences. A new Forest Act came into force in 1917 which came to be known as Forest Act I of 1326 Fasli with a view to overcome several problems faced by the Act of 1900.[40]

---

34 *The Evening Mail*, 2.6.1903 and *Meezan*, 7.10.1944.

35 Government of Andhra Pradesh, *Andhra Pradesh District Gazetteers,* Adilabad District (1976), p. 15.

36 HEH the Nizam's government, *Revenue Administration Report,* p. 130.

37 Ibid.

38 Ibid. p. 128.

39 Ibid. pp. 13–14.

40 Ibid., p. 14; also in Government of Hyderabad, *Questionnaire,* p. 20.

Forest Act No. I of 1326 Fasli (1917)

The Forest Act I of 1326 Fasli not only repealed the Forest Act II of 1900 but also was turned out to be more comprehensive than the earlier one. It contained all major provisions of the earlier Act and continued the three types of forests, i.e., Reserved, Protected and Open forests. A large area of the state had been placed under some sort of preservation under the provisions of Forest Act II of 1900. But in the case of reservation of forest areas, the Forest Department was miserably failed, according to the provisions of the Act of 1900.[41] Hence provisions were made in the new Act empowering the government to declare any such areas as reserved and protected forests. In these areas any rights of government or private persons over any land and forest produce had been inquired into, settled or recorded in a manner which the local government officials thought sufficient; or would be thus inquired into, settled and recorded after the declaration, if the former enquiries and settlement were considered insufficient. According to chapter II of this the reserved forests were the only forests par excellence because they alone were properly and fully secure. In this act all the essentials of forest constitution were observed; the boundaries were legally and strictly determined and demarcated on the ground; all rights of every kind were enquired into, recorded and authoritatively defined and provided for; all other rights not recorded were declared extinguished and no new rights were allowed to grow by prescription.

The reserved forests thus secured a permanent settlement whereas in protected forests existing rights were recorded but not settled, under chapter IV. Originally it was thought that a limited protection would be sufficient to the lands which could not at once be decided as reserve forests. This would save them from being cleared away and devastated. Under this Act, in the chapter III provision was made for the constitution of village forests. The thinking behind forming village forests was that village communities and private persons had a right to expect that government should provide protection to their forest property; protection against trespass and damage and also against the accrual of new rights. The village forests were to be under the management of government officer. It enabled government to take up any wasteland that was not reserved forest and at the disposal of the government to make it into a village forest. These regulations contained a number of rules for the management of village forests. The surplus revenue from the village forests was decided to go to the village community. But it

---

41 HEH the Nizam's government, *Revenue Administration Report,* p. 130.

was not clear who held the actual proprietary rights over this class of forests, the state or peasantry. Hence the concept of village forests could not be implemented successfully or seriously.[42]

The implementation of other provisions of the Act was a blow to the freedom of local communities where they had little or no access to reserved and protected forests and had very little in open forests. It was thought that the forests were often burdened with rights adverse to the public interest and hence they had to be dealt with in a special way. Certain acts like trespass and pasturing of cattle were prohibited in the reserved forests. These provisions gave right to the government to notify any forest or land as protected forests after ascertaining the nature and extent of the rights of the government and of private persons in and over the forestland or wasteland comprised therein. They were also empowered to make rules to regulate and prohibit certain acts in protected forests. Provisions were also made for providing suitable punishment for breaking of the forest law. The Act, on the other hand, empowered the Forest Officer to arrest any person against whom a reasonable suspicion existed of committing certain type of forest offence, without orders from the Magistrate and without warrant. Thus the Forest Act I of 1917 continued and extended the government policy of extending control over forests.

The Forest Department issued some stamps and permits as a part of concessions to peasants and tribal communities in 1923, 1924 and in 1929,[43] which would provide access to forest products such as firewood and fodder. These were originally issued to be used as grazing permits by the Forest Department as a means of collecting revenue. These stamps were printed in eight denominations and Forest Permits were printed in three designs and used to permit the holder to collect a head-load, a bullock-load or cart-load of firewood. The head-load permit was introduced in 1923 and the bullock-load and cart-load in 1929. For subsistence users, these permits were granted free of cost on the condition that the produce to use only for domestic and agricultural purposes, and not for sale.

---

42 Regarding the village forests as well, the government had retained the authority to cancel or assign the rights to village community though these forests were to provide timber, other forest produce and pasture to the people. The maximum area of any village forest would be 20 per cent of the total village area; then if the rest of the area was brought under cultivation, the village forest would be 25 per cent of the occupied area. No minimum area had been fixed as there were numerous villages which possess no forest at all; see, *ARFD (1916–17),* p. 11.

43 Raza Ali Khan, *Hyderabad 400 Years (1591–1991),* Hyderabad, 1991, p. 158.

On the suggestion of L. Mason, IGF to Government of India, as the Forest Advisor to the Hyderabad State, the state government defined its general forest policy for the first time under Revenue Secretariat (Forest Branch) Circular No. 11, dated 14 June, 1934. The important features of this policy were:[44]

1. avoidance of premature disafforestation
2. expedition of reservation proceedings
3. re-inclusion into proposed reserves of such slices of previously excised woodlands as contained forests of any value and were not fit for permanent cultivation
4. reservation, in the forestless tracts, of isolated detached blocks (including *Rumnas* and *Babulbans*) and transfer of their control from the Revenue to the Forest Department
5. limitation of free grazing concession only in cases where the area under *gairan* fell below 10 per cent subject to non-availability of grazing in the un-occupied village areas (outside the *gairan*).

L. Mason visited the Hyderabad State again in 1938 and inspected several forest areas of the state. In his report he stressed the urgent need of reservation of all forests in the state. He specially pointed out that out of a total forest area of 9700 sq. miles, only 8.75 sq. miles was constituted into reserved forests on his visit to the state.[45] Accordingly, he submitted a number of recommendations for remedying unsatisfactory position particularly in the constitution of the state forests. The result was that in its report of 1938, the Forest Department supported its efforts and existence in the Hyderabad State and stated that:

> The State has to determine the requirements and needs of a country, demarcate (forest) areas available, fix the legal status, protect them against several injuries such as human beings, fires, cattle and others, exploit them either by area or volume or number of trees and utilise all forest products in the most efficient manner, by first supplying the national requirements and if there is a deficit, to import them from foreign countries or if there is surplus, to export them which may be termed, 'forest policy'. .... Since all these activities involve a technical knowledge of managing the property, the State delegates its powers to an agency called that is best suited to carry on all these technical works which is known as the Forest Department.'[46]

---

44 Government of Hyderabad, *Questionnaire,* p. 23.

45 HEH the Nizam's Government, *Note on the Progress Made by the Hyderabad Forest Department Since My Report as Forest Advisor in 1341 Fasli,* by L. Mason, Inspector General of Forests, Government of India, Government Central Press, Hyderabad, 1938, p. 1.

46 *ARFD (1937–38),* pp. 1–2.

Thus, the policy emphasized the commercial use of the forest as also its ecological value. As a result it paved the way for the regulation of rights and privileges of people over forests and its produce. In the same year a revised Forest Act and draft Forest Code were submitted to the government for consideration.

For a discussion in the Legislative Council a note with the outlines of the forest policy and principles of the forest department was sent by IGF in 1941. Some of the important features were:[47]

1. producing forest mentality and consciousness through advertisement, propaganda and education
2. conserving, maintaining and consciousness through the existing forest areas and working them with the greatest possible care and caution under technically sustained yield management by first and foremost supplying the local agricultural and domestic village requirements at cheap concession rates or free supplying the forest requirements within the state at suitable prices and exporting the surplus or special commodities to provinces and countries abroad for an inflowing of capital
3. industrialising forest resources for small-scale cottage and large-scale power driven industry, saw-milling and starting other industries such as lac manufacture
4. cheapening the costs of transport particularly by railways and roads from forest areas to forestless ones
5. increasing the percentage of forest areas by control over private forests and unoccupied culturable and unculturable waste lands under suitable legislation and putting them to proper and suitable use.

IGF stated that the activities of the Forest Department would, henceforth, tend towards industrialization of forest resources of the state in addition to the rural development and village upliftment. He himself agreed that the Forest Department was generally condemned by the general public as it was regulating and restricting the people for the free access to forest resources. While supporting his department's activities, he said that the forest resources could not be unlimited and must follow certain natural biotic laws for the growth and development. He thought it was necessary to infuse forest mentality and conscience among the people by advertisement, propaganda and education, as the forests were meant for the people and their progeny and it had to be protected by the people. The Forest Department would look

---

47 HEH the Nizam's government, Revenue Department (Forest Branch), Instalment No. 219, 220, File No. 131/86, Record No. 420/1350 Fasli, APSA, 1941.

forward, according to him, to the day when it would not be regarded as enemy of the people but their supporter and well wisher.[48]

Further, in 1945, S. A. Vahid, IGF, suggested the following provisions vis-à-vis the forest policy of the state:[49]

1. better management of the existing forests
2. extension of the present forest area to the minimum necessary.

Better management of the existing forests to include:

1. putting the existing forest areas on a proper basis by declaring them Reserved forests as speedily as possible
2. immediate preparation of up-to-date working plans
3. proper control over grazing of sheep and goats from all reserved forests; enhancement of grazing rates to encourage stall feeding and
4. extensive research in the methods of regenerating the forests artificially or naturally.

While summing up he recommended that the future forest policy would be aimed at:[50]

1. Defining the areas necessary for the preservation of the general climatic and physical condition of the country. These areas should be kept as forests.
2. Creating the minimum amount of forest necessary for the general well-being by extension and afforestation. This minimum to be

---

48 Unfortunately, that day had not yet come as their goodwill always remained on papers and not in action. The Forest department increased their control over forests further through its latest Act of 1355 Fasli (1946) on the basis of Indian Forest Act of 1927. If it had a real wish to ameliorate the condition of the people it would have curbed its own subordinates from accepting bribery and the encroachments on the State forests by the influential feudal lords. The jagirdars, deshpandes, deshmukhs, and other feudal communities manipulated village records of forest areas through patels and patwaris and occupied many forest and grazing lands. They started giving permission to people by taking *nazaranas* and *mamuls* and those who refused to pay made to several harassments. Even though the common people reared their cattle in government forests these landlords compounded their cattle on the pretext of entering into private patta lands and so on. In addition to them, forest guards and subordinate staff were constantly worrying people in the villages by searching their houses, cattle-stalls, etc., in quest of offences and when forest produce was found with them were taken into custody on suspicion and the subsequent enquiry caused much annoyance and agony to the people; and were inflicted with heavy fines and severe punishments. The writings of the leaders of the Communist Party and the contemporary newspaper reports illustrate the plight of people.

49 Government of Hyderabad, *Questionnaire*, pp. 23–25.

50 Ibid., p. 25.

declared between 20 to 25 per cent. New land should be acquired and afforested to attain this.

3. Satisfaction of the local population at non-competitive rates especially in the aboriginal areas of the state is more important than revenue. Village plantations should be created to fulfull the local demands.
4. After the above have been satisfied, the object of management should be the realization of the greatest possible revenue consistent with the principles of forest management.

The draft Forest Act which was submitted to the government in 1938 became a law in 1946 and this came to be known as Forest Act No. II of 1355 Fasli. This Act was based on the Indian Forest Act of 1927.[51] Except some minor exceptions, the Forest Act No. II of 1946 of the Hyderabad State was, word to word, the same as the Indian Forest Act of 1927.

Forest Act No. II of 1355 Fasli (1946)

The government gradually increased its control over the forests and by this way the Forest Department was strengthened from time to time with a view to regulate people's rights over forests and its produce. According to this Act, forests were continued to be classified into reserved, protected and village forests, as in Indian Forest Act of 1927. According to this Act, clearing forest and granting of *pattas* were strictly prohibited in the reserved areas. No person could claim a property right in forestland merely because he was a brought up there, or even if his forefathers had lived there for centuries. Nor did such people have any rights over forest produce. Section 3 of this Act say, 'the government may declare any land in respect of which forest rights vest in the government or the land the whole of any part of the forest-produce of which the government is entitled, as reserved forest...'. The above section also indicates that the common land i.e., the place where people used to live in and around forest areas was the property of the government, and that the latter was *ipso facto* entitled to the forest produce.

Provisions made in respect of the protected forests were the same as in the earlier Act. A number of provisions were made for the protection of forests over lands which were not the property of the government (sections 35–38). Related to protected forests the government was empowered to enforce laws over it and to regulate the following matters under the sections 29–34: the cutting, sawing, conversion and removal of trees and timber of protected forests, the collection and subjection to manufacturing process and

---

51 The Political Secretary, HEH the Nizam's government, Letter No. 7913, dated, 6.8.1938, Instalment No. 16/75U, Record No. 547/1351 Fasli, APSA, Hyderabad.

of forest produce, the granting of license to cut and remove trees for personal use and for the purpose of trade, the protection from fires, cutting of grass and pasturing of cattle in the forests and so on. There was a penalty for the violation of each section of the above provisions. Thus, almost all the provisions in this Act were made within the framework laid down in the Indian Forest Act of 1927. The same type of forest administration based on these Forest Acts continued up to 1948 with some minor modifications and alterations.

Different Perceptions over the Forest Policy

The formation of the Forest Department in 1867 and the forest policies it enunciated had an everlasting and far-reaching impact on the society of the Hyderabad State. At that time agrarian society of Hyderabad State had a very close dependence on forests. It was not only the forest dwelling tribal communities who were dependent on forests but the entire rural population whose main source of livelihood and income was dependent on forests in some way or the other. With the sustainable style of living, in the normal conditions poor peasants were dependent on forests for their domestic and agricultural needs i.e., timber, fuel and grazing. Artisans collected their raw material from forests. In case of natural calamities too, forests were the most important alternative to these communities as they could rely on them for their survival. Also they could migrate from one place to another in case of famine or drought or any other disaster.[52] However, with the establishment of the Forest Department, the poor people lost their supplementary source of livelihood. Rearing cattle, which was an essential component in the life of the cultivator, became difficult due to the restrictions imposed by the Forest Department on grazing. Artisans also lost their source of raw material from the forests due to government restrictions. This situation was not only a case of the Hyderabad State but these conditions prevailed all over India during the colonial regime.[53]

---

52 The study of the famines in detail in the Hyderabad State will be discussed in the sixth chapter. Dependence of people during famine periods was more prior to the establishment of the Forest Department. A study on this aspect during famine of 1897 in Bihar region is undertaken by Vinita Damodaran, 'Famine in a forest tract: Ecological change and the causes of the 1897 famine in Chota Nagpur, Northern India', in Richard Grove, V. Damodaran and S. Sangwan (eds.), *Nature and the Orient,* pp. 853–890. *Broken Promises: Popular Protest, Indian Nationalism and the Congress Party in Bihar, 1935–1946,* Oxford University Press, New Delhi, 1992.

53 For details on Madras Presidency, see A. Murali, 'Whose trees? Forest practices and local communities in Andhra, 1600–1922', in David Arnold and R. Guha (eds.), *Nature,*

The state forest policy in the form of Forest Acts resulted in increasing restrictions on the peasant and tribal communities and their use of forest-produce. Each successive Act increased state's control over forest areas which meant increasing restrictions of the people. More and more forestlands were converted into the state forests for revenue purposes or to meet the needs of British imperial at the cost of the subsistence needs of the people. It is clear from the above section that the Forest Acts and policies adopted right from the establishment of the Forest Department in the Hyderabad State were formulated and reformulated to enhance commercial use of forests and this largely ignored the needs of the peasants and the tribals.

The forest policy of the Hyderabad State made it clear that the over-riding consideration would be 'public interest', which could be interpreted as the interest of the larger economy; where it conflicted with local interest.[54] The integration of forests with the larger economy and the gradual inclusion of a large proportion of them under the state forests thus proved to be financially a success to the government.[55] Although it was officially declared that the policy behind the system of management of forests was 'to conserve, maintain and improve the existing forest areas and working them with the greatest possible care and caution under technically sustained yield management by first and foremost supply the local agricultural and domestic

---

Culture, Imperialism, pp. 86–122; for all India level impact by the British forest policies see, chapters in Part Two from 19 to 24 on 'Colonial forest management and its impact on indigenous societies' in Richard H. Grove, V. Damodaran and S. Sangwan, (eds.), *Nature and the Orient*.

54 The ideological premises and material interests of the Raj, connected with the all India British Forest Acts were closely adopted in the Hyderabad State. There was a general discontent amongst people in the State that the forest policy declared in 1890 was not implemented in its spirit. Otherwise it would have given prime importance to the local needs of the people. The destruction of forests for commercial purposes was justified in the name of 'public interest'; where as cultivation of arable lands shown as forestlands but without any actual tree cover was treated as encroachment. The revenue oriented policy of the Forest department was criticised by its sister department, i.e., the Revenue Department in its report of 1914–1915, that the occupation was the production of revenue with a minimum amount of protection; see *Revenue Administration Report 1914–1915,* pp. v and 135. In the general administration report of the Hyderabad State, Mr. Ballantine, the Conservator of Forests, criticised the department on his arrival in 1887, that the Forest department had been useful as a collecting agency and the matters relating to the conservation and reproduction it had failed; see, *Annual Administration of the HEH the Nizam's Dominions 1303 Fasli, (1893–94),* Bombay, 1895, p. 69.

55 The continuous increase of gross forest revenue from Rs. 1,09,354 in 1881 to Rs. 1,14,05,876 in 1945 shows the enormous commercial value of forests in the State.

village requirements...'.[56] There was little substance behind that populist rhetoric, for it remained only on the papers not in practice; and the policy did not spell out the organizational reform needed to safeguard local interests at least where they did conflict with public interest. It did not suggest any decentralization of control in the form of giving over minor forests to the village or community administration at local levels.[57] It hardly helped in making actual people-oriented forest management and the trend towards tightening of restrictions on local use and the alienation of the local population from forest development continued. There was always a demand for surrendering the customary rights of the local communities to the state whenever formulating a forest policy. Officially, it was the local population who destroyed the forests while the commercial outlook of the government and the capitalists as well as the influential class's exploitation of the resources were not seen as factors.

The contemporary newspapers like *The Deccan Post, The Hyderabad Chronicle, The Evening Mail, Meezan*, etc. reflected the opinion of the common people in the late nineteenth and early twentieth centuries. From the 1930s onwards an important role was played by the leaders of the Andhra Maha Sabha and the newspapers like *The Golkonda Patrika* began to highlight and bring into focus the general poverty and the issues associated with the use of forests.[58] Suravaram Pratapa Reddy, Editor of *The Golkonda Patrika,* showed the miserable condition of the people and the problems faced by them in procuring the forest resources. One can see that, since the enforcement of the Forest Acts and policies news regarding forests and the plight of the people in general was a regular feature in all the newspapers.

---

56 HEH the Nizam's government, A Forest Note sent by the Revenue Department to the Secretary, Constitutional Affairs in 1941, IR, Instalment No. 219 and 220, File No. 131/86, Rec. No. 420/1350 Fasli (1941).

57 In fact, *The Golkonda Patrika* requested the Nizam's government to establish 'forest panchayats' involving people living in the nearby villages of the forest areas for the development and preservation of State forests, by giving free access to the grazing and the fuel needs of the villagers in the open forests to reduce the pressure on the reserved and protected forests; see, CVL. Narsimha Rao, 'Details of the Forest Department in the Nizam's Dominions' (Telugu), *The Golkonda Patrika,* 9.5.1938, p. 5.

58 Andhra Maha Sabha was founded in 1930 by the people inspired by the nationalism under the leadership of Suravaram Pratapa Reddy and other like-minded people. This organization was to serve objectives of the socioeconomic reforms, upliftment of peasants, educating the public on the crucial issues of the day and communicating the views of the people to the Nizam's government, etc. See, D. Venkateshwara Rao, *The History of Telangana Peoples' Armed Struggle – 1946–1951*, p. 148.

These newspapers and the organizations like Andhra Maha Sabha kept informing the government about the people's grievances over the forest issues through its news items, editorials, comments, deliberations and resolutions. *The Evening Mail* while reviewing the report of the Forest Department for 1901–1902 commented that:

> Hearty cooperation of the ryots of the villages, situated in the forests and their vicinities, is the most essential factor in preserving the forests and this sympathy could not be enlisted without kind and sympathetic treatment from the Forest staff. They need not be treated like criminals for minor forest offences, as from times immemorial the ryots have been thinking the forests of their villages to be their property and it is only since the creation of Forest department that the rules for preservation have been passed from time to time. Rigid compliance with these rules and heavy fines and sentences, will only aggravate the peasant's sense of justice and may lead to indiscriminate use of opportunities....[59]

The Golkonda Patrika also realized the importance of forests in the agricultural economy and the agrarian society like in the Hyderabad State needs a better management of natural resources like forests. It agreed that more severe regulations might be necessary in order to protect the forests. For all other forest produce, particularly in the open forests, including grazing rights, it believed, there was no requirement of any prohibition or detailed system of fees and permit on forest areas forming and produce.[60] Though *Golkonda Patrika* recognized the importance of a Forest Department it did not appreciate the way in which people's age old rights over forests were denied by the Forest Department. It held that the greed of the Forest Department to make money through forest produce was the most important cause of all the trouble. Also it advised the Department to limit its role to purely technical and advisory matters.[61] The principal objection against the rules and regulations was that they involved constant supervision and petty interference by forest subordinates. In fact senior forest officers acknowledged that the situation during this period had reached such a stage that the prosecutions for theft and trespass were already assuming an unenviable proportion in a forest officer's daily work making him regarded more of a police officer than a friend of the people.[62]

---

59 *The Evening Mail,* 5.5.1904.

60 CVL. Narsimha Rao, 'Details of the Forest Department in Hyderabad State', *The Golkonda Patrika,* 9.5.1938.

61 Ibid.

62 Ibid.

The Andhra Maha Sabha agreed in principle over the need to conserve the forests. It passed a series of resolutions in its fourth conference held at Sirsilla in January 1936, under the leadership of Mandumala Narsing Rao. The following resolutions/demands were made and the participants requested the government to implement them immediately:[63]

1. Facilities should be made by the government for free supply of agricultural implements to the peasants.
2. Obstructions created by the Forest Department should be removed in free grazing and firewood supply.
3. The agricultural lands were not being provided with natural manures like cowdung. Due to the restrictions of the Forest Department the peasants were forced to use cowdung instead of firewood, which could be otherwise used as manure to the fields. Hence a request was made to allow people to use forest resources for their fuel needs.

The Sabha also requested the government to remove all obstacles in procuring forest needs of the people. Due to the implementation of strict forest regulations, people were forced to come at illicit agreements with subordinate staff of the Forest Department by giving bribes, etc. To overcome all these difficulties, the Sabha requested government to change the system.[64]

The Deccan Post expressed its views in its review reports on the Forest Department in and around AD 1900. It asserted that the British government, which initiated this process in India, had claimed a right to own or regulate forest produce from the middle of the nineteenth century. However, after the formal establishment of the Forest Department it gradually enhanced its revenue earning capacity and began to curtail people's rights. The newspapers accused the Forest Department of the Hyderabad State for following in the

---

[63] *The Golkonda Patrika,* 13.1.1936. Also see, Ch. Rajeshwara Rao, Arutla Ramachandra Reddy and Y. V. Krishna Rao, *Bhooswamya Vidhanam Raddukai Raithula Veerochita Tirugubatu,* Telugu., (Heroic peasants' struggle against the landlord system) Andhra Pradesh Communist Samithi, Vijayawada, 1973, p. 23. Commenting on the plight of the Telangana peasants, noted freedom fighter N. G. Ranga said that the people were living in hard conditions who do not know modern agricultural techniques and technological know-how. The government officials were very corrupted and justice was not assured to the common people. He said that poor people were suffering a lot due to non-availability of fuelwood with ease. He asked the government to provide free supply of fuelwood to the people. See, N. G. Ranga, *Economic Organisation of Indian Villages,* Bombay, no date, quoted by Yugapati Simha in 'Plight of Telangana peasants', *The Golkonda Patrika,* 20.1.1936.

[64] Report of the Andhra Maha Sabha, published in *The Golkonda Patrika,* 6.12.1937.

footsteps of the British for perpetrating a sudden and violent confiscation of the long cherished traditional rights of the people since its establishment in 1867.[65]

Imposition of rules over the use of forest resources naturally gave rise to a lot of discontent in the society.[66] The needs of the people directly came into conflict with the interests of the Forest Department.[67] Even F. A. Lodge, the IGF (1914–1921), had written that 'The forest department is generally unpopular both with other departments and with the public, because its aim and objects are not understood. Its most obvious action is the curtailment or abolition of various long existing practices, and it is accused of causing hardship to the cultivating class, the cattle owner and the poorer classes.'[68] The Forest Department started taking over forest areas even before the Forest Act II of 1900. Large tracts of forests were demarcated particularly after the passing of Forest Act I of 1917 and the local rights were settled either by the Forest Settlement Officer (FSO) or by the Revenue Department officials. At

---

65 *The Deccan Post,* 15.6.1900 and 14.12.1900.

66 Actually it happened everywhere in India, for the study in Madras Presidency, see A. Murali, 'Alluri Sitarama Raju and the Manyam Rebellion of 1922–24' in *Social Scientist,* vol. 131, April 1984 and David Arnold, 'Rebellion Hillmen: The Gudem Rampa Raisings, 1893–1924' in Ranajit Guha (ed.), *Subaltern Studies,* vol. I. New Delhi, 1982; see I. M. Saldanha, for Bombay Presidency, 'Colonial Forest Regulations and Collective Resistance: 19th Century Thana District', in Richard Grove, V. Damodaran and S. Sangwan (eds.), *Nature and Orient,* pp. 708–733; see Ramachandra Guha for all India level, *State Forestry and Social Conflict in British India*.

67 The implementations of forest rules and regulations disillusioned the peasants and the tribals alike. In this process the forest officials behaved with highhandedness caused severe inconvenience to the needs of people. The news paper *The Evening Mail* commented that the public experienced great oppression at the hands of the officials of the Forest department creating large-scale discontent in the society; see, *The Evening Mail,* 18.6.1904.

68 F. A. Lodge, *Notes of Forest Policy and Forest Management*, Secunderabad, no date, p. i. (Preface). According to him, he had written this note in the hope of removing some of the false ideas regarding a much maligned Department, and spreading some little knowledge of the necessity for its existence and the objects it seeks to attain. However, in theory everything looks good and coming to the implementation of the forest policy these ideals were turned down by giving priority to the commercial needs of the State. This can also be seen in the proceedings of The Hyderabad Forest Conference held in 1332 Fasli (1923); S. H. Hussain Khan, the Divisional Forest Officer, Medak Division, in his presentation on 'Forest Policy for the Dominions and Its Necessity', criticised that 'The capitalists and the influential classes who really dictate the policy of the Government and found out another means of satisfying their craving for forest clad areas and the process of destruction and exploitation continued...', in *Proceedings of The Hyderabad Forest Conference, 1332 Fasli (1923),* 1924, Hyderabad, p. 1.

the most fundamental level, the demarcation and fencing of large forestlands meant effective loss of control by the tribal people and to some extent by the peasants over their habitat – a control so necessary for the reproduction of their existence.[69] Rules of conservancy were more strictly applied after 1917. Deprivation of age-old rights over forestlands had demoralised the people. The forests in which tribal communities had been dwelling since time immemorial were suddenly alienated from them. Commercialization of forests had led the forest officers, timber merchants and contractors to enter into the forests. The traditional forest dwellers could see these timber merchants and contractors felling trees taking away the timber to market towns.[70] However, the same action was not permitted to them. There was a feeling among people that unless money was paid nothing could be obtained from the forests. Tribals were closely linked with the forests and hence they were the worst affected lot amongst the total population. It was the state intrusion which disturbed the symbiotic relation between the forests and the tribals who were reduced to the mere position of labourers of the Forest Department and the urban timber merchants and contractors. People also felt cheated when they saw government cutting trees to supply to Hyderabad and other towns in British India while the local villagers were prohibited from doing so.

Many a times the tribals were turned out of their *gudems* (hamlets) by the Forest Department in order to protect reserve forest areas. Though these poor people were offered compensation in terms of money it was nowhere close to what they lost in terms of cattle, members of their own community, apart from the land. They did not like the land that was offered to them for cultivation as an alternative was generally of inferior sort to them. Added to such offensives, the forest officials exhibited much high-handedness and

---

69 Tribal people were always accused of illicit felling, misuse of concessions, age-old harmful habits, and were held responsible for dwindling forest cover. But, one thing which was lost sight of is that tribals had always lived in forests and the problem of deforestation never occurred before the commercialisation of forest resources. Curtailment of rights and privileges of tribals on the one hand and the disappearance of forests on the other hand, had really left the tribals in a sorrowful plight. The worst affected class due to the denudation was that of tribals, which is the most neglected one.

70 The tribal population particularly in the Godavary river tracts, due to the restrictions by the Forest department were not only deprived of the most ordinary chances of survival, but also subjected to unscrupulous exploitation by a group of timber merchants from the Madras Presidency who enjoyed monopoly rights over the extraction of forest produce; see, CVF. Haimendorf, 'Tribal Populations of Hyderabad Yesterday and Today' pp. XXIV–XXV; *Tribal Hyderabad,* 1945; also see, Syed Khaja Abdul Gafoor, *Tribes and Tribal Welfare in Hyderabad,* p. 39.

injustice towards the tribal people in the implementation of the rules in the evacuation process. As a result many *Patta* lands held by the tribals had been included in the reserved forests without proper compensation paid. The result was that the owners in many cases still paying the revenue for lands they were no longer allowed to cultivate.[71] In the case of Rudrangi village, in the *taluq* of Sirsilla of Karimnagar district, the cultivated lands of this village were situated in forest area of Bemugal Range of Nizamabad division. Many of these lands were held as *pattas* and some other lands cultivated by – Lambadis in the survey numbers of 1–17. The Forest Department came down heavily on these innocent people and asked them to vacate these cultivated lands as they were laid in forest area. They suggested opting for other lands available in other parts of the village or in any other village of their choice. Taking the advantage of people's illiteracy, the forest guard of the village took the thumb impression signatures of the villagers on a paper by not informing what was written on it. Later the Forest Department sent them a notice to vacate the land. All their efforts to petition the higher authorities went in vain. In this process neither the compensation was paid to the people nor they were not allotted new lands.[72]

If that is the plight of the tribal people, the peasants were also robbed of rights and grazing facilities that they enjoyed from times immemorial for the reservation of forest areas. All forest areas and many of *bids*, *porambokes*, *gairans*, and other waste and grazing lands were constituted as state forests leaving people almost deprived of their traditional facilities. Like grazing and collection of firewoods. If they were found dwelling in these forests it was considered as an act of offence. After selecting an area for reservation, the FSO notified that area under the provisions of Forest Acts and invited any claimants having claims on the proposed forest block to come forward. Claimants, if any, would submit their claims which were settled by the FSO and he would propose to declare the land as reserved.[73] In addition to it,

71 CVF. Haimendorf, *Tribal Hyderabad,* p. 92; *The Golkonda Patrika,* 17.7.1941.

72 *The Golkonda Patrika,* 17.7.1941. Actually this was not the case of Rudrangi alone and this type of incidents happened all over the State.

73 For example, Ozampur forest block, situated in Nalgonda district, was proposed to be reserved under section 18 of Forest Act I of 1326 Fasli. It had the area of 1914 acres and 7 guntas. This block was notified under section 7 in 1943 and the FSO under section 10 invited any claimants in 1943 having claims on the block to come forward. Seven cases were submitted and settled. The boundary of the block was finally demarcated and a map was attached to this declaration. General remarks were given at the end of the declaration in the following way, viz, 1. Sufficient area has been kept for *gairan* (for grazing purposes) for all adjoining villages, 2. Neither PWD Road, cart-track nor footpath passes

domestic animals were held up whenever they entered into state forests. The forest guards troubled people for trivial matters such as crossing reserved forests to reach their agricultural fields.[74] Bribing the forest guards to secure their connivance at getting their ordinary requirements in the forest produce was a regular phenomenon in the Hyderabad State. If more valuable timber was required something extra had to be given by the individual who wanted it. The poor who were being harassed by the revenue and forest staff naturally developed animosity against the authority of the forest officers.[75]

Another important change was that the government also tried to claim property rights for the common grazing lands and forests. The grazing rules came into existence in 1894 and revised in 1912, only to be made more burdensome. It also continued to be very unpopular among the people. In contrast to the relatively free use of wastelands for grazing purposes up to the middle of the nineteenth century, the present regulations were found to be confusing and irksome. Though, the state had, in some areas, allowed the villagers to utilize the village forests, loss of community ownership effectively broke the link between the man and the forest. The harmony of village communities was rudely shaken with de-recognition of communal rights and introduction of money economy, particularly after the establishment of Forest Department.

---

through the block (sometimes the previous cart-tracks pass through the blocks and rights to use these are admitted), 3. Villages are sufficiently far away to avoid any inconvenience to any of them, 4. The nearest block is Suryapet situated at a distance of one mile, and 5. Ranger's headquarters is situated at a distance of 12 miles but the forest guard lives quite close. Later opinion of *Subedar* (Medak Subah) was taken which was in favour of reservation. Consequently the FSO proposed to declare the block as reserved under section 18 of Forest Act. HEH the Nizam's government, the Revenue Department (Forests Branch), File No. 13/98, Record No. 111, 1356 Fasli, 1947, (IR).

74 Though officially people were allowed to use the roads crossing the reserved forests but were objected by the forest guards who used to take bribes as an advantage of the ignorance of the people; see *The Evening Mail,* 2.6.1903 and 18.6.1904, *Golkonda Patrika,* 5.8.1931 and 14.7.1938,

75 Some of their anger was reflected in setting up the fire in forest areas and some times rebelled against the Forest and Revenue authorities. For example, grazing fee was introduced in around 1895 and since then some of the officers of the Forest Department seemed to have shown certain energy in levying it But the peasants, graziers like *Dhangars,* and the tribals had shown even greater energy in turning their herds and cattle into the State forests and denuding them as goats and cattle so well know how to do; see, *The Hyderabad Chronicle,* 2.2.1898. In later period, this situation was changed slightly by going edge towards the Forest Department. The tribals, particularly the Gonds, rebelled against the atrocities of the Forest Department in 1940; see, *The Golkonda Patrika,* 2.9.1940 and 16.10.1940.

Instances of breach of forest laws, organization of public meetings to demand the lowering of land assessment, withdrawal of restrictions pertaining to the forest, etc. occurred frequently. The media continued to highlight such people's problems concerning forest issues. The torturous rules and regulations of the Forest Department and their implementation by the most ruthless forest officials continued to be a subject of criticism. Many issues of The *Golkonda Patrika* demanded to remove the grazing fees and advocate free access to forest products. It also appealed to the people to organize themselves and protest collectively against the unjust activities of the Forest Department.[76]

Poor peasants and tribals though discontented were in no position to protest in an organized manner. Individual offence was one form they could adhere to. Due to this the number of forest offences increased to such an extent that the Forest Department was identified as the police department by the common people.[77] Bribery and corruption among forest officials at the lower level became a common phenomenon. The most notorious practice was a blackmailing system in which forest guards used to threaten poor villagers with fines or imprisonment for imaginary violations of new regulations. The government cautioned the Forest Department to pay due regards to the habits and needs of the poorest class of the population, knowing well, that this class resented vexatious and oppressive interference with their daily life. However, the Forest Department did not pay heed to it.[78]

---

[76] The principal objection against the rules and regulations was that they involved constant supervision and petty interference by forest subordinates. In fact senior Forest Officers acknowledged that the situation during this period had reached such a stage that the prosecutions for theft and trespass were already assuming an unenviable proportion in a Forest Officer's daily work making him regarded more of a police officer than a friend of the people. In almost every month the newspapers and conferences of the Sabha received a large numbers of petitions of grievances from the rural population, some of them common to nearly all the districts of the State. These newspapers and the resolutions passed by the Sabha ventilated the grievances of the people and made the government to know their condition in rural and forested areas. They vehemently criticised the department for indulging in massive confiscation of land rights since 1900 and for the way corruption had grown in the Forest Department. The sudden imposition of restrictions on the use of forests and forest produce, he pointed out, had reduced the people to a choice between starvation and poverty. The Golkonda Patrika, 5.8.1931, 25.8.1938, 3.7.1941, 7.7.1941, 17.7.1941, 15.9.1941, etc.

[77] There were 5,139 cases were registered in 1901–1902 which rose to 7,423 in 1939–1940; see *ARFD,* 1901–1902, p. 11 and *ARFD,* 1939–1940, p. 13.

[78] *The Evening Mail* gave one of such examples in its issue of 23.8.1904, it described such as, 'The public experiences great oppression at the hands of the officials of the Forest department and is put to a severe loss. Particularly in Warangal district, the Forest

However, the discontent among the people over the forest issues continued to increase. Even after protecting the reserved areas, rest of the forested land that was meant for the use of people was not totally free in the true sense of the word. People's movements in these forests were also restricted. They could avail of the facility granted by the Forest Department only at certain times. Also this could be done only if people had permits and if they paid the grazing fees.[79] The regulation to prevent forest fire was the most controversial as it was difficult to discover who had set fire. The department not only required villagers to help to extinguish fire but they were held collectively responsible for fire if the guilty was not found. The newspapers like *The Hyderabad Chronicle* in its editorial commented that 'It is not so much by the application of penalties and by hard treatment that it will eventually be succeeded, but trying to make the people understand its value, and by constant watchfulness, good-protection-belts and the application of the rule that it is better to prevent people setting fire than to punish them for doing it. We are glad to see that the Board hopes that this important part of the forest conservancy will gradually be dealt with in an effective manner.'[80] For example, the fire in Nirmal Reserve in 1902, which was not controlled well by the Forest Department, was attributed to incendiarism of villagers. The Assistant Conservator of Forests suggested the suspension of the rights

department did not sanction *naulowni* (pattas on forest area) application contrary to government circular. They are not allowed to cultivate uncultivated lands by cutting down bushes growing there on, although this was authorised by the kowls they hold. The Forest officials do not pay attention to the kowls or the orders of the government.' Though the Forest Department admitted the existence of maladministration and corruption in the Department in its annual administrative reports and did very little in this regard. It gives a list of punishments such as reductions in the grade, dismissals and suspensions of the staff. But in its report of 1948–1949, it suggested, 'Punishments should tend more towards constructing and reclaiming subordinates than destroying them which was generally observed in giving punishments.'

79 *The Evening Mail,* 11.11.1899 and 18.6.1904; *The Golkonda Patrika,* 5.8.1931, 25.8.1938, 17.7.1941, 21.8.1941; *Meezan,* 26.9.1945. This type of oppression prevailed in all parts of the State and some of them cited above were given in the newspapers.

80 *The Hyderabad Chronicle,* 18.6.1898. Fire was a great enemy to the Hyderabad forests as to the forests of British India. But establishment was quite inadequate in the Hyderabad State and the average area of each guard's beat was 50 square miles or 5 times that in the Central Provinces. The protection of forests from fires and browsing by sheep and goats was reported by the Conservator of Forests and the measures taken were quite insufficient and particularly the open forests appeared to be rapidly deteriorating. The contemporary opinion was also reflected the same thing that it was hopeless to attempt conservation worthy of the name; see *Ibid.,* 2.2.1898 and *The Deccan Post,* 6.8.1900.

of the villagers who did not turn out to help in extinguishing fire within their village limits. Though his suggestions were heeded by the Conservator of Forests, it shows the high-handedness of the staff.[81]

Thus, during the study period there was a concern to develop a professional body and forest legislations in support of the implementation of forest policy. The whole unfolding of policy was built on economic concepts favouring higher efficiency, increased control over the people and the resource, and centralization of power. In the initial stages the preservation of existing forests became the *raison d'être* of the Forest Department. In the beginning the survey and demarcation of forestlands were carried out with the view to conserve forests. Later, the whole policy of conservation was pursued with the object of facilitating production of timber, both for revenue and for British imperial purposes. However, the Hyderabad State was not an exceptional one in this issue.

It is clear that the forest policy of the Hyderabad State gave maximum importance to the revenue aspect of forests and commercialized supply of major forest produce in order to comply with the general forest needs of the state. It also implemented in the form of Forest Acts resulted in taking over of forests and forest produce. The policy adopted by the Hyderabad State was in continuation of the British colonial revenue-oriented approach, that converted the forests into a supplier of raw-material for the forest based industries and a source of revenue for the government. The priority of the state forestry was essentially commercial in nature. This trend was due to the financial crisis that ruined the state in the middle of the nineteenth century. Naturally all the efforts of Salarjung I were aimed at generating more revenue to the state. The work of the Forest Department was not an exception. Its activities were judged on the basis of the revenue it generated. The state's takeover of forests and the subsequent working of these on commercial lines had the curtailment of customary rights. The loss of forests and pastures, earlier communally managed and enjoyed, severely undermined the subsistence economy of the peasant. This led to the animosity in the part of the community towards the protection and management of forests.

---

[81] *ARFD,* 1901–1902, p. 12.

CHAPTER 4

Forest Administration and Management

As a part of reforms in the state administration taken up by Salarjung I, all matters connected with forests were delegated to the newly created Forest Department in 1867. Thus, a foundation was laid down for a 'systematic' development of forestry in the Hyderabad State. The Forest Department was under the control of Home Secretariat until 1881, and it was later handed over to the Revenue Department.

Structure of Administration

The Conservator of Forests was the head of the Forest Department. The department was reorganized in 1914–1915 and the post of Conservator of Forests was elevated to the rank of Inspector General of Forests (IGF) with two Conservators under him. Under the IGF, there was a hierarchy of staff consisting of Conservators, Assistant Conservators, Rangers, Deputy Rangers, Foresters, Surveyors and Guards.

The forest areas of the Hyderabad State were divided into circles, divisions, ranges, blocks and beats. In 1899, there were four forest divisions consisting of Warangal, Bidar, Gulbarga and Aurangabad. In 1901 they were increased to six with the creation of Mahabubnagar and Nirmal. In 1914–1915 the divisions were increased to 12 and were grouped into two circles, styled as Eastern and Western circles, each under the charge of a Conservator.

This system continued until 1948.[1] Each division was looked after by a Divisional Forest Officer (DFO) and the work of the DFO was controlled and supervised by the Inspector General of Forests. Each Divisional Forest Officer had a number of graded staff of assistants called *Ameens/Darogas* (Rangers), *Sahradars* (Foresters), *Dafadars* (Assistant Foresters) and *Chowkidars* (Guards). During the reorganization of the department in 1914–1915, two posts of Working Plan Officers were created.

The following table 1 shows the number, grades and pay of subordinates as well as the head of the Forest Department during 1914–1915.

Table 1. Forest administration set-up in Hyderabad State

Designation	Qualifications	Number	Initial pay	Full-grade pay
Inspector General of Forests	European training and experience	1	1500	1500
Conservator of Forests	Do	2	1200	1200
Assistant Conservator of Forests	1. European training and provincial service course 2. Either conditions 3. Do	16 (in total)	350 250 250	1000 600 600
Sub–Assistant Conservator of Forests	Trained in a recognized school	8	150	200
Rangers	Do	68	70	130
Deputy Rangers	Do	50	35	70
Foresters	Trained locally	86	25	35
Surveyors	1. Grade A 2. Grade B	18 (in total)	80 45	154 85
Forest Guards	–	720	12	15

Source: Report on the Administration of HEH the Nizam's Dominions for 1331 Fasli (1921–22), Companion Volume, p. 78.

In addition to the above mentioned staff, there was one full-time Settlement Officer, Working Plan Officers and other officers who were temporarily appointed and generally deputed from the Revenue Department. There were, as well, office clerks and cashiers and other servants, not shown in the above table. However, the department in the Hyderabad State was very

[1] *Report on the Working of the Departments under the Director-General of Revenue, 1324 Fasli (1914–15)*, pp. 9–10.

small and was also somewhat low paid compared to the British Provinces of similar size in forest areas. For instance, there were only 6 Assistant Conservators in Hyderabad State compared to 24 in Central Provinces, 18 Rangers to 124, and 43 Foresters to 180 and 374 Guards to 1471. In total, there were only 442 staff members of all cadres in Hyderabad State compared to 1801 in Central Provinces. Moreover, remuneration was very low in the Hyderabad State.[2]

The number and low remuneration indicate how small was the Forest Department in the State compared to the contemporary standards. This establishment was consistently unable to ensure the Government a proper understanding of its organizational needs. The department truly lacked qualified foresters, and did not have sufficient number of subordinate staff to work and protect the forest property, to regulate grazing according to rules and to carry out the improvements decided upon. The Foresters repeatedly requested the government in their annual reports to increase the pay and the establishment.[3]

Training and Education

D. Brandis, the first Inspector General of Forests to the Government of India, recognized the need for a qualified and scientifically trained personnel in forestry for the management, administration and conservation of forests in India. With the increased demand for trained staff at various levels to implement the forest policies, several forest schools were established in India at several places to meet the requirement starting with the one at Dehradun in 1878.[4]

In Hyderabad State, after 1880s, the regular staff trained in forestry service began to be recruited. The IGF or Conservator of Forests was recruited generally from the Imperial Forest Service personnel or the persons trained in Indian Forest Service. The DFOs were taken from the candidates who passed out from the forest training schools of British India, or, who after long service had proved themselves fit for the position. For efficient working of the department in technical matters, suitable candidates were sent out for

---

2 For instance, Conservator's salary range was Rs.1200 in Hyderabad State, compared to Rs. 1300–1800 in CP, also Ranger's salary was only Rs. 50–80, compared to Rs. 40–150 and Forest Guard's salary was Rs. 7–8 compared to Rs. 6–12. *ARFD,* (1900–1911), p. 27.

3 Ibid; and L. Mason, *Note on the,* p. 7; *ARFD* (1937–1938), p. 4.

4 For a detailed study of the forestry education in British India of Imperial Forest Service, etc., see the chapter on 'Forestry education: History and development' in *100 Years of Indian Forestry – 1861–1961,* Vol. II, FRI, Dehradun, 1961, pp. 105–112.

training in British India and also in Europe at government expense. They were employed in higher scales on their return.[5] The subordinate services like Rangers were generally recruited at local levels. For the recruitment of Rangers, local training classes were conducted from time to time. There was no state institution until the establishment of a forest school for the training of forest staff particularly Rangers in 1916 at Nizamabad. It was, however, closed down after giving training to only one batch of Rangers. Afterwards, the candidates of Conservators were sent for training to the Imperial Forest Training College at Dehradun and Rangers to Madras Forest College, Coimbatore. Later a Forest Training School was established in the State in 1929 at Mahabubabad to train Deputy Rangers, Foresters, Forest Guards and lower grade staff.[6] However, only permanent staff and candidates for Forest Guards and Foresters were given training mainly in this school.

There were two sessions separately maintained in this school to train literate and illiterate Forest Guards. Illiterate ones were only given practical training for six months. For the literate Guards, after training and practical sessions an examination was conducted in which they had to pass out successfully. In addition to them, classes were also conducted for Rangers, Deputy Rangers and Foresters. The Forest School at Mahabubabad was shifted to Yellendu in Khammam district in 1940. It was established there permanently, as the conditions and facilities were satisfactory. An industrial centre was also established at Yellendu where the students were taught a practical outlook on forest industries.[7]

Numerous publications, forest records and leaflets from the Forest Research Institute, Dehradun were obtained and a number of modern books on forest literature were purchased for the benefit of the Forest Department staff. The Forest Department library maintained books on modern forestry and catalogued up to date all the information on forestry and management. The department started publishing a magazine called *Hyderabad Forest Magazine* from 1940 onwards which was exclusively devoted to the latest developments in forestry. Thus the Forest Department brought forest and other officers in the State as well as educated common people to know about forest matters.[8]

---

5 *ARFD,* (1893–1894), p. 72.

6 *ARFD,* (1938–1889), pp. 50–51.

7 HEH the Nizam's government, the Revenue Department, Review on the work of the Forest Department for 1347 Fasli (1937–1938), Instalment No. 19, File No. 1/30, Record No. 547/51F, APSA.

8 ARFD, (1937–1938), p. 39.

Forest Management in the Hyderabad State

Forest Department began its work on the adherence of the principles of forest conservancy and declared that the State was going to take over all the forests that was not proved to be the property of private individuals. However, for about two decades after its establishment the Forest Department simply maintained and protected some species of trees, designated as 'reserved' (*Irsali*) and the activity of the department consisted of preserving these trees, whereas the rest of the species and produce remained under the control of the Revenue Department. The effect of this dual control over forest tracts was that even many large timber-producing areas were cleared for agrarian expansion.[9] The real value of forests was underestimated, as was clearly exhibited by the provision of many of *ijara* settlement throughout the State, which transformed large forest tracts forever into agricultural lands. Added to this, Forest Department staff was inadequate both in numbers and training. By the late nineteenth century, Forest Department was convinced that revenue and village officials could not understand the need for long-range conservation and systematic application of the forest regulations. The department began pressing to have professional foresters and trained forest guards to replace the unskilled men in this work.

Declaration of 3390 sq. miles of area in the year 1894 as protected forest and placing it under the sole charge of the Forest Department marked the commencement of a new era in the forest management of the Hyderabad State. Grazing rules and regulations were framed enabling foresters to exclude cattle from forests. Moreover, cattle were divided into household and commercial categories. The classification of the cattle was at the discretion of the forest officials. While officials preferred stall-feeding/fodder-cutting to grazing, cattle owners preferred to take the cattle to grazing grounds than to bring the fodder to the cattle. Though the Hyderabad State followed the national agricultural policy, based on J. F. Voelcker's recommendations for the creation of fodder reserves for cattle, however, very little was done regarding this matter. Further, cattle were largely seen as a hindrance to successful forestry. Goats and sheep were considered as foes of the forest.

---

9 The Revenue Department considered forests as an obstruction to agricultural development and no apprehension was felt that the supply of forest produce would ever fall short of things would last forever. As a result forests were treated as an obstacle to the prosperity of the State and the watchword of the time was to bring everywhere more forest tracts under cultivation, and the whole policy was tended in that direction. *Questionnaire of the Forest Policy Committee,* Government of Hyderabad, Hyderabad, 1956, p. 20.

The Forest Department excluded even some areas from sheep and goat grazing, for 'regeneration' purposes.

Forest Act II was passed in 1900 to reserve forests in Telangana districts where large-scale deforestation for cultivation and colonization was under contemplation. With this Act, the State acquired complete legal control over forests. Exploration and demarcation of the forest areas were followed by efforts to introduce protection and regular management, working plans, and survey schemes, being main plans of work. In addition to enunciating ground rules on felling, Forest Department also advocated regular tending and pruning. Unfortunately large forest tracts had already been laid bare in Warangal and Karimnagar districts and were given away for cultivation to Banjars, Ijaras and Maqtas.

In 1914, F. A. Lodge, IGF, prepared a Forest Code in 1921 and laid foundation for the systematic exploitation of some forests and reservation of important forest tracts. In the year 1917, the Forest Act II of 1900 was amended laying down detailed procedure for reservation and for controlling the incidence of forest offences. Like their British counterparts, the Forest Department in Hyderabad State was expected to evaluate the magnitude of the disturbances to the forest caused by both nature and humans. It was expected to take measures to restore the equilibrium of growth and yield over time. Forests across the State began to be surveyed and inventories tracts of forest plant species made. Working plans and silvicultural operations were also devised and implemented for the entire State along the guidelines of the British forestry.[10] Following the practices of British India, Lodge worked for the replacement of felled trees by sowing seeds or planting seedlings in the gaps, landings and other forest openings. Another aspect of the guidelines for management enunciated by Brandis for Indian forests was the mitigation of factors that would be inimical to forest growth, such as grazing, fire and vicious weeds. This was also followed-up by Lodge in the Hyderabad State. Communications were improved and infrastructure for management of forests, such as roads and buildings were constructed.

---

[10] Brandis's silvicultural guidelines were calculated explicitly to prevent either under or over-utilization of wood. In his first working plan for Pegu, he stipulated that the exploitation of forests in a given period should not exceed their increment. The annual yield, he thus laid down, should be based on the amount of growth, they put on. This principle subsequently became the cornerstone of all Indian forest Working Plans. M. Buchy, 'Quest for a sustainable forest management: A study of the working plans of North Canara district (1890–1945)', in Ajay Rawat, *Indian Forestry: A Perspective,* Indus Publishing Company, New Delhi, 1993, pp. 141–162.

After Lodge retired in 1921, the department came under the control of Nawab Hamid Yar Jung Bahadur, a non-technical officer, for a period of nearly seventeen years. This period was very critical for the Forest Department, because the contradiction between preserving the forests and meeting the people's needs was becoming acute at the time. The demands of forest work always exceeded their resources. During this period the Development Department was created (in 1922) with the object of establishing colonies for persons from densely populated areas of the State, and also for immigrants.[11] This department was kept directly under the control of Sadre-Azam Bahadur (President, Executive Council). The energies of Forest Department were spent in convincing the government of the necessity of forests in the economy of a country and pressing for a change in its *Anti-Forest Policy* i.e., clearing of forest lands for agriculture and colonization. As a result of vigorous protests of Forest Department, the colonization scheme was abandoned.

L. Mason, the IGF to the Government of India, acted as the Forest Adviser in 1932. He visited important forest areas of the State and gave technical advice and instructions on the management of forest resource. Based on his re-commendations, systematic surveys were undertaken. This streamlined the marketing of forest products. The protected forests, which were hitherto not surveyed for the character of the vegetation or for the pattern of users' rights, were studied and reclassified as reserved rorests.[12] Simultaneously, controls over the day-to-day use of the forests were attempted, and this strict process left the foresters with few friends but many critics.

In continuation of Mason's recommendations, forest survey and settlement operations were embarked upon during the late 1930s and most of the protected forests were turned into reserved forests. Preparations of working plans according to recent advances in tropical forestry were started. In order

---

[11] Rural Reconstruction Secretariat, HEH the Nizam's government, File No. 139/6 and S. No. 2 of 1358 Fasli (1949).

[12] The process of reservation of forest lands was complicated by the fact that many forest tracts were intercepted by private estates, patta lands and private forests. The ownership pattern was too fragmented for effective land management, and the Department faced many problems in survey and settlement, demarcation, and in litigation in its attempts to consolidate forest tracts before they could be effectively reserved and managed. In this process, many poor and illiterate people were not given due compensation and put to many hardships. This process, as elsewhere in British India, produced lengthy disputes and friction between the Forest Department and the people. HEH the Nizam's government, Revenue Department, Instalment No. 219, File No. 131/86, Record No. 420/1350 Fasli, 1941, APSA, Hyderabad.

to organize and systematize the preparation of working plans, the IGF convened many conferences of Conservators and working plan officers. In 1938, it was resolved that division-wise working plans had to be prepared by combining all identical types of forests under the same method of treatment and revising the plans in the light of the above considerations.[13] Reclassification of timber in accordance with the new ideas about utility of various timbers was also brought into force. Separate schedule of rates for different zones were fixed and worked out. The Forest Department sent proposals to the government for the rules regarding the restriction of free grants and prohibition of grazing by sheep and enhancement of grazing fees, which were sanctioned in 1944. Except areas under regeneration, and those that were prohibited grazing, usually grazing was allowed in the reserved and protected forests. General instructions with regard to demarcation, cases of forest offences, auction sales and fire protection were issued in the same year.

Considering the general utility of the produce and yield as well as forest types, the total forest area of 830/sq. miles under the control of the Forest Department was broadly classified in 1941 as follows:[14]

1. timber forests (big and small) – 4,000 sq. miles
2. scrub forests, including fuel forests, grazing grounds and grass lands – 2,000 sq. miles and
3. hill forests (rather unproductive and protection forests) – 2,301 sq. miles.

Realizing the unsystematic over-exploitation and reduction in density of forest tracts, the Forest Department took up conserving, maintaining and improving the existing forest areas and working them with the 'greatest possible care and caution under technically sustained yield management'. While passing the remarks over the utilization of forest resources, the IGF stated, 'We cannot grow any species un-saleable to the locality.'[15] This shows that despite lofty ideals, ultimately the Forest Department like its British counterpart showed interest only in the species which had commercial value.

---

13 ARFD, (1937–38), pp. 7–8.

14 HEH the Nizam's government, Revenue Department, Circular, Installment No. 219, File No. 131/86, Record No. 420/1350 Fasli, 1941, APSA.

15 'Brief notes on forest utilization', by Conservator of Forests, Eastern Circle, Instalment No. 220, File No. 96/86, Record No. 330/50, (IRH), 1941.

In a note of 1945, the Revenue Department announced to the State Legislature the *raison detre* of the Forest Department[16], which consisted of the following important points:

1. to reserve all forest areas suitable for conservation and protect against several injuries such as human beings, fires, cattle and others
2. to increase forest area by artificial regeneration and afforestation and formation of fuel and fodder reserves
3. to improve the quality and quantity of grass in the interest of the cattle that form the real backbone and motive power of every agricultural activity within the State
4. to bring inaccessible big timber and pole forest crops under systematic working
5. to stop soil erosion and disintegration by increasing the forest density and filling up gaps
6. to provide the forest dwellers and the local agricultural population within the state with all possible preferential privileges and concession for their domestic, agricultural and industrial requirements at the cheapest possible rates
7. to utilize soft and fancy timbers for the start of handicrafts, small cottage and large power driven forest industries
8. to export the surplus, if any, to foreign countries creating demand and markets outside
9. to protect games and preserve wildlife by formation of shooting blocks, game preserves and game sanctuaries.

The lofty ideals and progress of the Forest Department were stated regularly in its annual administrative reports. According to it, despite meeting all the local demands for the timber, fuel, agricultural and domestic requirements as well as grazing and grass requirements of innumerable men and cattle, the department was a source of good income to the state and justified its operations and mechanism.[17]

---

[16] *Policy and Programs of the Forest Department,* Revenue Department, Instalment No. 219 and 220, File No. 131/86, Record No. 420/1355 Fasli, Interim Repository, Hyderabad.

[17] Though the income from the Forest Department was around 2 per cent of the total income to the State exchequer, it was of minor significance when considered individually. However, with the other sources of income it collectively occupied a very significant place in the revenue complex of the State, which accounted for nearly 30 per cent of the total revenue. B. K. Narayan, *Finances and Fiscal,* p. 65.

Technical Management of Forests

In the management of the forest and its resources, prior to its exploitation certain assessments were carried out relating to survey and settlement of the forests, creation of forest reserves, demarcation of the boundaries of the forests, consolidation of its demarcated areas, and mapping of the forest resources.

Survey and Settlement

This was an important aspect of State forestry because the State control and timber extraction of a forest tract was based on the control of the land, through the implementation of the forest settlement legalized by the Forest Acts. The idea of demarcating the forest areas into blocks and divisions emerged in the late 1880s but was not properly implemented till 1914–1915.[18] A Forest Settlement Officer (FSO) was appointed in 1914–1915 for this purpose, till then this work was done by Assistant Conservator of forests or by the revenue officials deputed for this purpose. In addition to this post, the services of T. E. M. Cladius, a retired officer of the Survey of India, had been secured for forest survey work in the state.[19] By the 1920s the forest settlement became a systematic body with clear objectives and procedures of application precisely defined by the Forest Act of 1917 and by 1930 the survey of state forests was completed.[20]

The work of survey and settlement was classed under three headings:

1. demarcation of grazing and arable lands for the several villages situated in or around the State forests;
2. demarcating the boundaries of the State forests themselves and
3. preparing a survey of the forests so demarcated.[21]

---

18 HEH the Nizam's government, Forest Department, Instalment No. 38, List No. 2, Private Secretary's Office, File No. 13, G2/C1, June, 24, 1882. R. S. Dobbs, Conservator of Forests, recommended for immediate action, in his own words, '...that this is work which should not be lightly undertaken and further, if not carefully done it would be worse than useless as it would give rise to numberless quarrels and disputes'. However, all such types of recommendations went futile till the year 1914–1915 due to official negligence.

19 *Report on the Working of the Departments under the Director General of Revenue, 1324 Fasli (1914–15),* no date, p. 2.

20 HEH the Nizam's government, Revenue Department, a letter sent to the Political Secretary, for the perusal by the Director, Geodetic Branch, Survey of India, Dehradun, Letter No. 1481, 27.6.1940.

21 Forest Department File No. 65/97, Record No. 10/56 Fasli, Instalment No. 233, (IRH), 1944.

The aim and purpose of the settlement of forest tracts was twofold: first aim was to ascertain and spell out the extent of the state's proprietary rights over forest lands and clearings, as well as over forest produce, as defined by the Forest Acts. The second aim was to record and inquire about the existence of eventual private rights which interfered with those of the state and, subsequently, to negotiate the prerogative of the latter to ensure that forest tracts would be managed for the public benefit and for future generations.[22]

Demarcation and Mapping

According to the law in the state, every immovable property, private or state should be demarcated to enable it to be identified by the general public, so that in case it was trespassed, proper and due action may be taken. 'The only effectual remedy is to demarcate the most promising forest tracts, and close them against all interference.'[23] The Forest Department of the Hyderabad State agreed to this comment made by D. Brandis. For this purpose, the department cut a boundary line, 15′ to 30′ wide with earth, stone or wooden pillars at suitable intervals. By 1930s almost all the forest blocks which were under the charge of the Forest Department had a clear boundary line cut between them the neighbouring property. [24] A DFO had to keep a record of reserved, protected or open forests, in the form of a map. Though conducting survey of all the forest blocks in a division was not possible at one time. Therefore, a rough map was drawn up from the local settlement maps and kept for convenience sake. [25]

Working Plans

A working plan is a technical arrangement under which the growing stock, accumulated in forest areas, is not only kept intact and maintained, but also improved and added to, whilst the annual increment is extracted every year so that a sustained yield and revenue may be secured.[26] It was through the working plans that the state's forest policy was translated into action. Within the framework of the general forest policy of the state, a working plan was prepared for producing a sustained of the most needed forest products, in the

---

22 Ibid., pp. 7–9.

23 Quoted in Ajay Skaria, 'Timber conservancy, desiccationaism and scientific forestry: The Dangs 1840s–1920s', in R. Grove, *et al.*, (ed.), *Nature and the Orient*, pp. 596–635,

24 ARFD, (1937–38), p. 6.

25 HEH the Nizam's government, Revenue Department, a letter sent to the Political Secretary, for the perusal by the Director, Geodetic Branch, Survey of India, Dehradun, Letter No. 1481, 27.6.1940.

26 *Revenue Administration Report*, (1914–15), pp. 134–35.

most profitable manner, and then making them available to the people conveniently and at reasonable cost. The working plans were aimed at systematic long-term intervention in the use of government forests, with a view to raising productivity. They secured protection of several tree species that were important sources of revenue. Prior to its formulation, access to most state forests had been on commutation terms. Access to all parts of a reserve was permitted once a fee had been paid to the Forest Department. Till F. A. Lodge's term as IGF (1914–1921), there was no systematic extraction of forest produce. In the reorganization of the Forest Department during 1914–1915, two full time working plan officers' posts were created each one for Eastern and Western Circles. The objective was to evaluate the annual growth of a stand and to extract there from each year without jeopardizing the future growth of the forest.[27]

But until 1930s, no regular working plans were implemented in the State. Commenting on the situation, Mr L. Mason, the Forest Adviser to the Hyderabad State said: 'A properly prepared working plan is the first essential for the efficient management of the forests and it is, therefore, fundamental that this branch of the department should, without any delay, be put on a satisfactory footing.'[28] The first working plan was prepared in the year 1938 for Nirmal forest division under 'Coppice-with-Reserves' system.[29] From 1936 to 1948 two full time working plan officers started preparing working plans for various divisions and a separate working plan circle was formulated in 1946. The work of preparation of working plans had been systematized in 1940 in the light of modern scientific ecology, industrial development and protection of wildlife and preservation fauna.

Silviculture System of Management

Silviculture was really synonymous with forestry; but as regards technical instruction, the term was limited to that branch of forest science which deals with the formation, treatment and reproduction of woods.[30] A large part of the forests in the Hyderabad State were managed under the working plan

---

27 ARFD, (1937–1938), p. 7.

28 L. Mason, *Note on the progress,* p. 7; ARFD, (1937–1938), p. 4.

29 *Questionnaire of the Forest Policy Committee,* Government of Hyderabad, Hyderabad, 1956, p. 26.

30 NAI, Hyderabad Residency Records, Political Branch (1832–1948), Serial No. 708 – F-249, 1946. And John Nisbet, *The Forester: A Practical Treatise on British Forestry and Arboriculture for Land Owners, Land Agents and Foresters,* Vol. I, William Blackwood and sons, Edinburgh and London, 1905, pp. 62–63.

concept, involving a number of silvicultural systems. Another strategy in the implementation of the working plans, aimed at achieving the economic objectives of the state, was to choose the most appropriate methods of silviculture.[31]

Various coppice systems, which were characterized by short rotations and regeneration of the trees through sprouting from cut stumps instead of seed, were used primarily in the Hyderabad State. Silvicultural operations were undertaken in the State according to prescriptions laid down in the various working plans in force, though not much had been done owing to lack of proper protection and absence of means of communication. As the large number of forest tracts were under-stocked often this led to the deterioration of the soil. This should be pointed out that thinning operations were not permissible in those tracts. Planting up blanks was undertaken by sowing seeds of hard species. First and foremost consideration of a working plan officer, according to Forest Department, was to render crop normal, which was not so easy in the State, because the subordinate staff were not well trained in working plans and silvicultural operations.[32]

Forests were looked upon, under these silvicultural practices, as a source of revenue and not as a means of meeting the genuine needs of the forestry. That is why the thrust of forestry had been for the high forest system, which called for felling or cutting back all growth except the species chosen for dominance. The main system adopted was 'Coppice with Standard' where it was desirable to get large-sized poles, in addition to 'Simple Coppice'. There was hardly any area under selection system in the Hyderabad State. All these different systems prescribed methods of treatment, suitable to the locality for which they were meant under the different objects of management of producing timber, fuel or grazing.[33]

Methods of Regeneration

Natural Regeneration

Forest Department kept some exploited areas aside and protected them from grazing and fire, for natural regeneration by seeds as well as of coppice shoots. In some other areas under the natural regeneration, coppice shoots, pollarded shoots and seedlings were flourished well on arrival of rains during monsoon. Most of the natural seedlings of teak, satin, nallamaddi and other

---

[31] M. H. Ali, Working Plan Officer, Western Circle, Instalment No. 265, File No. 6/86, Record No. 1093/1338 Fasli, 1927 (APSA, Hyderabad).

[32] Ibid.

[33] Ibid.

species, that died back in previous years, would shoot up. All the annual reports of the Forest Department reported the natural regeneration and on the whole the results in the State were shown as satisfactory.

Artificial Regeneration

The replacement of trees, extracted or minor products removed, was one of most important technical duties of the Forest Department. In the Hyderabad State where the value of forest produce was rather low, 'it was much more economical, rather cheap, to have recourse to and depend upon natural regeneration, but whenever it failed, for want of a good seed bed or absence of seed trees, to go in for artificial regeneration: certain special methods of regeneration such as *taungya* and *rab* were involved in the country.'[34]

Rab regeneration was attempted in Khammam, Karimnagar, Mahabubnagar, Asifabad, Warangal and Nanded divisions on an area of 165 acres. *Taungya* plantation was done in Asifabad, Mahabubnagar, Nanded, Warangal and Aurangabad divisions on an area of 83 acres. Leaf manure plantations, mostly of *karanj* and *tephrosia-candida* in Nizamabad and Nalgonda were carried out on an area of 305 acres. In Himayathsagar and Jidmitla, nurseries of green leaf manure, *Casuarina* and avenue trees were also set up. *Sandal* was raised by dibbling under bushes in patches in Aurangabad and Gulbarga. Stump planting of teak was done in Warangal, Karimnagar, Asifabad, Nirmal, Nalgonda and Mahabubnagar divisions. In other areas, broadcasting and pit sowings or planting were attempted. The plantations of Medak division, i.e., Gandipet, Himayathsagar and Jidmitla adjoining Hyderabad city, were too small and were therefore converted to forest park, experimental farm, plantation and forest research and experimental station respectively. Wet nurseries were established in almost all the ranges in each division.[35]

Under a scheme of reafforestation the soil and moisture were conserved by reclamation operations. The soil thus conserved had to be strengthened by introduction of grasses and shrubs. The Forest Department did some experimental work on rather a very small scale in Shorapur, Gulbarga district and in some hill areas of Raichur district. Only 29 acres of area were afforested in Shorapur due to financial problems. It is expected that 3 inches

34 The *taungya* system was nothing but an imitation of the agricultural system where *tuar* was sown in lines and cotton or other crops were sown in between. The *rab* system was also an imitation of agricultural methods, in which forest slush, just as used by the agriculturists in their paddy fields, was collected in strips, or patches, heaped up and burnt. This served as a good seed and germination bed for artificial as well as natural regeneration. ARFD, (1937–1938), p. 16.

35 Ibid, p. 17.

of soil would be accumulated every year in the trenches. *Siris* was found to be a fast growing species and *Cassia siamea* and other species were introduced as suggested by L. Mason, Inspector General of Forests, to Government of India.[36] Soil reclamation operations were also started in the Himayatsagar experimental farm under terracing, trenching and pitting.

Research and Experiment

These activities were taken up in the State only in 1930s. Under the Resolutions passed at the 4th Silvicultural Conference held at Dehradun in 1934, preservation plots in Asifabad and Karimnagar divisions were protected and maintained and measurements were recorded periodically. Experiments on seed origin were started in 1938 with seeds from Rangoon, Mysore and Malabar in Nizamabad, Mahabubnagar and Medak divisions.[37] The exploitation of nallamaddi, bijasal and mohwa species for railway sleepers had been started in the 1890s on experimental basis.[38] As the teak was very expensive and available in limited quantities, research was on to find out other species suitable for railway sleepers. During the same period, experiments were going on at Dehradun on the species of chir and blue pine.[39] Of all the timbers, nallamaddi and bijasal were found suitable in Hyderabad State and available in large quantities and could be marketed at a sufficiently low price for railway sleepers.

Communications and Buildings

Construction of motor roads made the forest resources more accessible for commercial exploitation. However, till the 1920s there were no good forest roads constructed in the State and little money was spent annually on the improvement of existing cart tracks which served as line of export for forest produce.[40] Forest roads led to exploitation of the forest resources to the fullest extent. Gradually in all the state forests extensive roads were constructed to exploit the forest resources quickly and efficiently. All the permanent buildings were constructed for the Forest Department and repaired by the Public Works Department. After the 1930s, all administrative reports of the Forest Department gave the details of buildings constructed or repaired with the expenditure incurred.

---

36 L. Mason, *Note on the progress,* pp. 9–12; ARFD, (1937–1938), pp. 18–19.

37 ARFD, (1937–1938), p. 19.

38 GAD, (1894–1895), p. 12.

39 R. S. Pearson, 'Note on the antiseptic treatment of timber in India, with special reference to railway sleepers', *Indian Forest Records,* vol. II, pt I. Calcutta, 1918; quoted by Ramachandra Guha, *The Unquiet Woods,* pp. 43–44 and 47.

40 L. Mason, *Note on the,* pp. 6 and 8.

Conservation Measures

Until the beginning of the twentieth century, conservation measures were not taken up seriously. The jurisdiction of the Forest Department was confined to control over 16 species of trees and all other species were under the control of the Revenue Department, except some small forest reserves.[41] This system was proved destructive to the existence of forests as the cultivation was unnecessarily permitted by the revenue officers in the forest areas. Under the cloak of clearing land for cultivation, much valuable timber was destroyed. A system of giving free grants of timber was also abused and the Forest Department was unable to exercise any effectual check on it.[42] Grazing too was also not restricted; according to the Conservator of Forests, 'millions of trees' were destroyed due to unrestricted grazing. The system of granting forest lands under *ijara* and colonization scheme for the development of agriculture continued till the first decade of the twentieth century. The Revenue Department's development policy under these schemes had been largely responsible for deforestation in the late nineteenth and the beginning of the twentieth century. After the assumption of control by F. A. Lodge, as the Inspector General of Forests, conservation measures were taken up by the Forest Department.

According to the departmental report most of the forest areas in the Hyderabad State, particularly those in the vicinity of villages, were subject to over-felling, and were under-stocked, except the inaccessible parts. The hilly areas and reverine areas of Mulug, Khammam, Madhira, Yellendu, Paloncha, Mahadeopur, Chinnur, Utnur and Nirmal in Telangana region were prone to the danger of erosion whenever heavy rains occurred. This happened due to the sloppy and undulating topography of the land. Remedial measures like regulated forestry, controlled grazing, regeneration, etc. were thought to be given the attention of the Forest Department.[43]

---

41 HEH the Nizam's government, Forest Department, Instalment No. 38, List No. 2, Private Secretary's Office, File No. 6, 9.8.1878; and ARFD, (1900–01), p.7.

42 Ibid., File No. 13, G2/C1. A letter from Forest Department' office to the Minister, R. S. Dobbs, Conservator of Forests during 1880–87, proposed for reforms in the department regarding power division between the Forest Department and the Revenue Department and showed how the misuse of power and free-grants given to the ryots. He strongly recommended for full powers to be given to the Forest department for forest conservation.

43 *Andhra Pradesh District Gazetteers,* Warangal district (1976), p. 72; Khammam district (1977), p. 54; Karimnagar district (1974), p. 84 and Adilabad district (1976), p. 55.

Fire Protection

The important and the most difficult aspect of forest management was to combat the centuries old custom of burning the forests for pastures. During summers the peasants and graziers resorted to the fire to obtain a fresh crop of grass. Although mature teak trees were remarkably resistant to fire, continuous reproduction in areas of commercial logging operations rendered the growth of young trees particularly vulnerable to fire. Fires in the forests were very common and very detrimental to the forest growth. The forest areas of the Hyderabad State were subjected to incessant fires for years, leading to a general unsoundness of the growing stock. Universal fire protection was a difficult task and a source of complaint for the peasants who were not accustomed to any interference with their customary practices. However, the State undertook measures to minimize the conditions that would cause fire in areas under management.[44]

The early initiatives for preventing fires and excluding livestock from state forests laid the ground for a wider regime of control under forest regulations since 1880s. Fire protection in the Hyderabad State attempted first time in 1897.[45] Coupe lines usually served for fire lines. These were annually, before summer, burnt under supervision and their width varied in different localities – from 4 feet to 10 feet. The Forest Acts gave them the executive powers of strict control on felling, tapping, lopping, girdling and burning forests. Several aspects of subsistence practices came under siege with the strict implementation of fire protection works. Several features of people's activities were labelled as 'illegal' or 'incendiarism' by the Forest Department. Gradually the area under fire protection increased and in later years rigid supervision and control over the large areas proved a greater success in percentage of the total area under fire-protection. The Forest Department also started a system of giving rewards to the aboriginal tribes who successfully protected and of withholding them if failure was observed. And whenever tribals cleared land for temporary cultivation by burning trees,

---

44 *Notes Regarding Administration of His Highness the Nizam's the Forest Department, From 1867 to 1911 AD,* quoted by M. H. Ali, Working Plan Officer, Western Circle, Instalment No. 265, File No. 6/86, Record No. 1093/1338 Fasli, 1927.

45 Brandis, Inspector General of Forests to the Government of India, elevated the prevention of fire from a local tactic to a 'leading principle' of forestry. He advised forest officials to control fires only in areas with a 'promise of good production of timber and other produce'; NAI, Revenue and Agriculture Department, Forests, September 1876, A. nos 15–19, pp. 3–4. The Forest Department of the Hyderabad State adopted and followed these measures in the State forests from the last decade of the nineteenth century.

the Forest Department saw fire as the fatal foe of timber trees and tried to stop this practice with stringent measures. However, it wanted to carry out a give and take policy by giving some privileges and rewards in case of success and punishment in case of 'malicious fires'.[46]

Grazing and Lopping

The strict exclusion of fire and grazing from state forests was a main feature of forest management in the Hyderabad State. According to the Forest Department the most important injury, to which the forest areas were liable, was unregulated and unrestricted grazing in areas, opened out to grazing, as well as illicit grazing in closed areas under regeneration. Almost all officials favoured the exclusion of livestock in order to assist the regeneration of important timber producing trees like teak, bizasal and nallamaddi. Prohibition of grazing was considered as the only way of improving the teak re-growth. Till the beginning of the twentieth century the Forest Department was not much under pressure to allow grazing in vast areas. The matter was not merely how many livestock were allowed into forests, it was closely bound up with the assessment of the impact of cattle on the vegetation in general and timber trees in particular. Though the Forest Department had a strong bias against livestock, it was not practical to impose complete ban even in reserved forests. Officials were in favour of fodder cutting as against cattle grazing in the forest lands. They felt that cattle caused considerable damage to young saplings, while fodder-cutting was less harmful. According to Forest Department, goats and sheep were the most 'powerful' or universal cause of deforestation. It was argued that goats were indiscriminate browsers and the entry of herdsmen increased the risk of fires. It prohibited stripping the bark of trees, lopping the twigs and the collection of leaves. The Forest Department charged peasants and herdsmen a grazing fee for allowing their animals' access to forests.[47] Again differentiation was made between browsers and grazers; and prescriptions were laid down in view of it in working plans.[48]

---

46 L. Mason, *Note on the progress,* pp. 13–14.

47 *Notes Regarding Administration of His Highness the Nizam's Forest Department, From 1867 to 1911 AD,* quoted by M. H. Ali, Working Plan Officer, Western Circle, Instalment No. 265, File No. 6/86, Record No. 1093/1338 Fasli, 1927.

48 The exclusion of cattle and sheep became a general policy of the Forest Department. It was critical of granting permission to grazing in the closed areas for regeneration; and held that sheep was much more harmful to forests than cattle. Sheep damaged trees and young saplings and graziers often set fire to the forest to get the fresh crop of grass. The Forest Department continued to charge graziers to discourage the entry into forested lands, which led to the discontent among them.

However, the lower officials of the department made use of grazing regulations and started harassing local communities by impounding cattle on real and fabricated charges. The methods of showing discontent such as the breaching of rules concerning cutting of trees and grazing, the burning of the forest floor for a fresh crop of grass, etc. were in effect as assertion of traditional rights whose exercise was now hampered by the implementation of scientific forestry. Naturally, commercial forestry divided the forests into blocks, which were completely closed after the trees were felled, and disrupted existing patterns of resource utilization. In order to enable regeneration to take place in the logged areas, closure to humans and cattle was essential. Likewise, in the areas under reproduction fire was to be avoided as dangerous to young saplings. Thus the continuance of traditional use of forest resources represented a major threat to commercial exploitation of forests. This was especially true for the firing of the forest, in contravention of the forest regulations, which sometimes resulted in a total annihilation of young regeneration.

Impact of the World Wars

Both the First and Second World Wars drove deforestation in the Hyderabad State, as elsewhere in India, as huge demands for timber forced the modification or abandonment of working plans to achieve higher production. 'The World Wars saw India's forests being requisitioned for the imperial cause. In fact, it was only during the "Great War" that the enormous potential value of India's forests for the Empire was fully recognised.'[49] They made heavy demands of timber, particularly poles. 'The working plan prescriptions were honoured more in their breach.' In many valuable forests, felling had to be carried out in total disregard of the working plan prescriptions. During the First World War there was an urgent demand for timber, which led to its large-scale cutting and extraction. However, the impact of First World War was not great on Hyderabad forests, if compared to the impact of Second World War.

The impact of the Second World War was more severely felt on the Indian forests. In 1940, a Timber Directorate was set up in Delhi to channel supplies of forest produce from the provinces. India was the sole supplier of timber to the Middle Eastern region, and later on to the Allied Forces in Iraq and the Persian Gulf.[50] During the Second World War most of the forests

---

49 E. P. Stebbing, *The Forests of India,* vol. I, p. 36, quoted by Ramachandra Guha, 'Forestry in British and Post-British India', pp. 1882–1896, especially p. 1887.

50 '...The available evidence points to considerable over-felling in forests of both private owners and native states, as well as to unrecorded felling in government forests – all to

under systematic working in the Hyderabad State, had to be overlooked in order to cope up with the War supplies. The demand for timber was of a much greater magnitude than that during the First World War.[51] Enormous quantities of timber of almost every species were extracted necessitating excessive felling and advance working in almost all forests of the State.[52] While exact figures are unobtainable, the extraction of timber was considerable, especially during the Second World War. Revenue of the Forest Department rose from Rs. 14,01,230 in 1939–40 to Rs. 1,14,05,876 in 1944–45. The situation in private forests and other princely states along with the British Indian forests was not different.[53]

Ecological Degradation

The State possessed a very inadequate and badly distributed area of forest, a large proportion of which was un-productive and the remaining was considerably impoverished due to lack of adequate conservation measures. Desert conditions were created in Marathwada and Karnataka regions with alarming rapidity and were gradually spreading eastwards. The most important natural injury in the State, especially in areas of low rainfall, sparsely clad with forest, was soil erosion. The scrub forest areas in the surroundings of Hyderabad and also in Marathwada, Karnataka and also in forest-less blocks of Telangana were being eroded year after year and a lot of soil was being lost.[54] In a working plan report of the Pakhal forest reserve in Warangal

meet the war demand.' Ramachandra Guha, 'Forestry in British and Post-British India', pp. 1882–1896, especially p. 1887.

51 Recommendations of Board of Forestry 1942, HEH the Nizam's government, Office of Executive Council, File No. 549, G2/C11, 20.4.1943, APSA; and *Questionnaire of the Forest Policy Committee,* Government of Hyderabad, Hyderabad, 1956, pp. 26 and 31.

52 NAI, Hyderabad Residency Records, Political Branch (1832–1948), Serial No. 713, F–270 – P, 1946, Note on a tour of inspection of Hyderabad forests by Mr A. P. F. Hamilton, IGF, Government of India.

53 Statistical Year Book, 1351–1354 Fasli (1941–42 to 44–45), pp. 637–649. For the details of other native states and British Indian Provinces, see Government of India, Ministry of Agriculture, 'India's forests and the war', Delhi, 1948; E. P. Stebbing's *Forests of India,* vol. IV, Oxford, 1962, Chapter 9.

54 The Karnataka and the Marathwada districts were much below minimum which had a percentage varying from zero in Osmanabad to a maximum of 6.20 in Nanded; Nalgonda and Nizamabad and Medak in Telangana districts were much below minimum. In the hilly areas of Marathwada, Karnatak and some Telangana districts repeated extraction of forest growth without giving a rest, goat browsing and unrestricted grazing of innumerable cattle had led the area bare of forest growth and eventually denudation and erosion had stripped off the soil and sometimes sub-soil exposing bare sheet rocks. ARFD, (1938–39), p. 22.

district, Inspector General of Forests, Mr Lodge in a letter to his Private Secretary commented on the degradation of the forest tract and the failure of the Forest Department and the policy in the protection of forests. He mentioned: 'The only result of the above policy has been that the growing stock and the soil have very much deteriorated, the density of the forest has considerably decreased, natural regeneration is entirely wanting and it will take at least 80 years to restore the forest to its normal condition.'[55] It was noticed in the working plans that such areas were protected and reclaimed before any clear felling were attempted.

A glance at the figures showed that the total area under forests in the State was only 11.4 per cent. It was accepted and declared policy of the Government of India applicable to the states as well, that the proportion of forest to agriculture should be 20 to 25 per cent. To make up this deficiency the Forest Department made abortive attempts to include the private forests and the holders of these private forests were proved too powerful to subdue to the action of the Forest Department. Later the department started taking up uncultivable wastelands for re-afforestation as protective measures against erosion and denudation.[56] To stop erosion and conserve both soil and moisture to a certain extent, *Anona squamosa* (sitaphal) and *Cassia auriculata* (tarward) seeds were sown in some scrub forest areas under and not under working plans and clear felling (simple coppice system) were gradually done away within such areas. For insolation and drought, pits within forest areas were filled up with slush and leaves, etc., and mulching was tried in some plantations.[57] For example, the Forest Department spent Rs. 1,48,620 on conservancy works in 1935–36, which indicated better progress in the improvement and development of forest areas.

Commercial Exploitation of Forests

The process of commercialization of forests started when the timber traders and the State began to use the forest resources for the larger economy particularly in respect of major produce.[58] Commercialization brought about a

---

[55] HEH the Nizam's government, Private Secretary's Office, Instalment No. 38, List No. 38, File No. 369, G1/C270, 11.10.1919, APSA.

[56] Policy and programmes of the Forest Department submitted by the Inspector General of Forests to the Revenue Department, Instalment No. 219 and 220, File No. 131/86, Record No. 420/1350 Fasli, 1941, (IRH).

[57] ARFD, (1939–40), p. 14.

[58] For example, forest produce had only limited commercial importance till the British started laying an extensive network of railway lines using wooden sleepers. Between 1853 and 1910, 51,658 km of track was laid which meant increase in the revenue of the

change in the perception of the forests and forests were no longer considered as inexhaustible. Commercialization ended the perception of abundance of forest resources and brought in the feeling of scarcity. Outside interests began to assert their control over forests and impose restrictions on local use. For this, the need for a systematic management of forests began to be felt both to regulate local use and to ensure regeneration of commercially more valuable forest produce. As a result of this the State took over forests and started regulating the forest resources under the newly established department through various legal acts.

The Forest Acts of the Hyderabad State provided the underpinnings for the 'scientific' management of forests, enabling the working of compact blocks of forest for commercial timber production. The State was able to maintain strict control over forest utilization and with the legal and institutional structure, the Forest Department extended its control over the forest lands of 5,186.2 sq. miles in 1900–1901 to 9,516.87 sq. miles in 1940 in the category of reserved, protected and open forests.[59] The Forest Department had to generate adequate revenue, adopting the method of British policy, so that each administrative machinery had to be self-supporting. Thus, the main efforts of the Forest Department were directed towards the commercial exploitation of the forests, because its means were limited. In addition to this the government considered the forests as a source of revenue. The objective of most of the working plans was to extract timber and other forest produce for sale.

Though timber and other forest produce have been used since ancient times, forest produce gained considerable commercial importance only after the establishment of Forest Department. This was reflected in the State's forest policy, in which forests became a source of revenue.[60] The basis of exploitation of forest was that it was treated as a form of capital not as a natural resource. Exploitation, therefore, was done on commercial lines to increase the revenue collection. Issuing the following communiqué to the Nizam's government and other British provinces by the Department of Education, Health and Lands, Government of India clearly stated that, 'The forests of India represent capital resources of great value. Much valuable

---

Forest Department from Rs. 0.05 millions to Rs. 239.90 millions for the same period. *History of Indian Railways,* Government of India, Delhi, 1964, p. 214.

59 ARFD, (1939–40), p. 3.

60 In the year 1891, the Secretary to the government clearly stated that, 'it is against the standing rules of H.H. government to supply timber at free of cost from the government forests.' while replying to an applicant; File No. 22, G1/C2, 1891, APSA.

work has been done at the Forest Research Institute, Dehradun, to test the practical possibilities of timber and other forest products for commercial and industrial use.'[61]

Agency of Exploitation

The methods of extraction and disposal of forest products were worked under two different systems, viz, departmental system and contract system. Under the departmental system the timber or other forest produce was brought from the forests to timber depots or other places of sale on account of the government. Under the contract (also called license or permit) system, the produce was sold in the forest and removed by the purchasers. In forests worked by departmental agency an account was kept for the timber or other produce removed from the forests, but this was not always practicable under the contract system.[62]

In the early years, before there was adequate establishment in the Forest Department, merchants were given full rights to timber harvesting on specified tracts for specified periods of time. The merchants then hired day labour and chose which trees to cut. This system devastated forest resources and knowing the danger of the system, the department began exercising the right to designate which trees to cut in any timber concession. In enunciating the approach of the British to silviculture, Forest Department laid down strict rules that governed felling. Contractors engaged in felling had to follow the minimum felling diameter set by the Forest Department. The process of felling had to be supervised by professional foresters on the ground, to ensure that forests were not destroyed.[63] However, the merchants began to evade the new controls, in pursuit of maximum short-run spoils. They adopted several strategies, most of them illegal, in their pursuit of profits for their selfish end.[64] Almost all traders who took permission for cutting timber

---

61 Government of India, Department of Education, Health and Lands, *Communique*, Simla, 15.5.1939, Instalment No. 38–2, File No. 546, APSA.

62 M. H. Ali, Working Plan Officer, Western Circle, Instalment No. 265, File No. 6/86, Record No. 1093/1338 Fasli, 1927.

63 GAD, 1303 Fasli, (1893–94), p. 72.

64 The forest policy of Hyderabad was determined largely by the interest and needs of the dominant groups in society. The commercial and industrial interests dictated forest policy at the expense of tribal communities and their life-support system. S. H. Hussain Khan, the DFO, Medak Division, in his presentation on 'Forest policy for the dominions and its necessity', criticized that 'The capitalists and the influential classes who really dictate the policy of the Government and found out another means of satisfying their craving for forest clad areas and the process of destruction and exploitation continued...', in

and forest products cut trees and clumps themselves, by employing local tribal and village communities, cut haphazardly and destructively, so that the Forest Department attempted to cut timber with their own labour, and hauled them to auction at timber depots or purchasers were expected to take timber.[65]

In the later period, timber was either sold by standing coupes by auction, or extracted by contractors or by the department itself, which directly hired labour and assembled the wood for sale at one of its timber depots. The departmental extraction was less destructive than contractors' extraction, but it could not be widely resorted to because of the paucity of available staff. The techniques of extraction were extremely wasteful on account of lack of expertise and inadequate and insufficient supervision.[66]

How the forests of the Hyderabad State were assigned the primary role of a revenue-generating organ by the Forest Department of the State can be seen from official records and documents. The following statement (Table 2) shows the overall decennial progress of the Forest Department.

Table 2. Revenue and expenditure of Forest Department (Decennial)

Year	Revenue Rs.	Expenditure Rs.	Surplus Rs.
1894–95	2,00,679	1,25,079	75,600
1904–05	6,30,962	2,61,918	3,69,044
1914–15	8,93,966	3,79,473	5,14,493
1924–25	16,10,763	6,67,806	3,42,957
1934–35	13,56,402	8,98,663	4,57,739
1944–45	1,14,05,876	25,48,766	88,57,110

Source: Compiled from *Administrative Reports of the Forest Department* of respective years.

Proceedings of The Hyderabad Forest Conference, 1332 Fasli, (1923), 1924, Government press, Hyderabad, p. 1.

65 The Forest Department marked specific trees to be felled by contractors who brought them for sale either in the local market or for British Indian markets. The local people had to secure their fuel and timber at non-competitive rates from the contractors. The economic life of peasantry was affected as they could not secure their forest requirements cheaply and easily. The contractors also illegally cut down more trees than marked and cheated both the government and the public for their illicit cuttings and higher prices. S. Kesava Iyengar, *Economic Investigations in the Hyderabad State, 1929–30, Warangal district, Vol. III,* Government central press, Hyderabad, 1932, p. 7.

66 'Brief Notes on Wood Technology', Conservator of Forests, Eastern Circle, 1939. Illegal transportation of Hyderabad timber into British territory was a common phenomenon in the State; and the lack of effective prevention of smuggling of teak incurred severe losses to the revenue of the Forest Department. *The Deccan Chronicle,* 4.3.1949.

As the above table indicates that the state management demonstrated the commercial value of the state forests and in common with other native chiefs, the Nizam became 'generally very much alive' to the value of his forest property. Forest expenditure usually formed not more than 50 per cent of revenue. If the figures given above are examined, they would indicate that the increase in the revenue ranged from 3 to 15 times, while the expenditure ranged only from 2 to 6 times. Surplus had accrued thus, from year to year. Forest exploitation reached the maximum during the Second World War period, which showed how the ideals of conservation were kept aside, if the British imperialism demanded.

For commercial purposes, the forest products of the Hyderabad State were divided into major and minor forest produce. Major forest produce consisted most of wood for use either as timber or fuel. Apart from wood, forests yielded a great variety of produce such as bamboo, bark (used considerably for tanning), gums, lac, honey, wax, canes, leaves, flowers, seeds, grass, etc. All these produce were described as minor forest produce (MFP). Frequently minor forest produce was of much greater value than the wood produced in the forests. Many of the minor products could be obtained from forests without disturbing the output of the major produce, but this did not happen in the case of Hyderabad.[67] Table 3 shows the lion's share of major produce out of the total out-turns of the forest products in the Hyderabad State.

Table 3. Revenue of forest products in Hyderabad State (in Rs.)

Year	Timber and fuelwood Rs.	Bamboo Rs.	Grazing/ Grass Rs.	Bark and leaves Rs.	Other minor produce Rs.	Miscella- neous Rs.	Total revenue Rs.
1895	1,53,913	–	11,443	–	–	33,990	2,00,346
1900	1,42,221	17,558	47,494	28,757	31,691	20,653	2,88,376
1905	2,87,577	22,644	1,16,335	10,826	–	83,700	5,21,082
1910	3,79,793	36,173	1,35,039	6,978	–	83,516	6,41,499
1915	4,27,434	25,950	3,53,721	8,994	34,843	54,062	8,93,966
1920	2,63,256	13,362	3,38,156	38,054	8,225	6,82,109	13,43,162
1925	6,46,756	39,402	–	74,966	7,46,888	3,34,441	18,42,453
1930	8,22,456	31,232	–	1,36,444	10,90,360	36,317	21,66,809
1935	7,13,577	29,003	4,40,889	79,126	–	93,805	13,56,400
1940	7,30,519	33,804	3,52,242	1,08,981	–	73,380	14,01,230
1945	68,56,600	208	10,72,313	27,16,627	96,132	3,07,553	1,12,57,284

Source: Compiled from *Annual Administrative Reports* of concerned years.

[67] A. I. Qureshi, *The Economic Development of Hyderabad,* vol. I, p. 14.

The above table shows the forest products under various heads which constituted the revenue of the Forest Department of the Hyderabad State. The Forest Department had mainly concentrated on the major produce, particularly timber, which had alone a share of more than 50 per cent of the total revenue. A large number of working plans or working schemes in the State were applied with the main objective of the extraction of teak species. Of all the timbers, teak was the main species which the Forest Department exploited most. It was in great demand and was extensively used earlier for ship-building and for construction of railway network since the middle of the nineteenth century. Timber and fuel formed the major source of revenue with 50 per cent of the total revenue of the Forest Department and was largely exported to the British provinces or in other parts of the State. Other species, such as nallamaddi, sandra, bijasal, etc. were largely consumed locally.[68] Timber requirements such as pit props for the Singareni coal mines at Bellampalli, Mandamarri and Ramakrishnapur in Adilabad district and Ramagundam in Karimnagar district, had been largely met from the forests of Adilabad. The Singareni collieries at Kothagudem in Khammam district needed about 24,000 pit props for its mines at Kothagudem, which were obtained from the local forests. Some of the forest coupes within a radius of about 32 kms from Kothagudem were taken by the Singareni collieries for pit props.[69]

Next to timber and fuel, grass and grazing charges formed a significant source of revenue to the Forest Department. Grazing in the reserved or protected forests was regulated by the Forest Department, and the fee was collected either departmentally or through contract agency. Grazing rights in the open forests were auctioned annually by the Revenue Department.[70] Thus, the peasant had to pay grazing tax (*Bancherai*) both to the Forest and Revenue Departments. However, the Forest Department paid attention mainly to timber producing areas by neglecting grazing grounds and grass lands.[71] The government *bancherai* was situated in many villages at a distance and there was no controlling authority to enforce steps to ensure proper growth of grass by preventing indiscriminate grazing. As such, it served the useful

---

[68] S. C. Dube, *Economic Life of Hyderabad,* Government Central Press, Hyderabad, 1937, p. 9.

[69] *Andhra Pradesh District Gazetteers, Khammam district,* Government Central Press, Hyderabad, 1977, pp. 15–16.

[70] *The Imperial Gazetteer of India, Hyderabad State,* p. 37.

[71] M. M. Ali Baig, 'Cattle improvement in Hyderabad', *The Hyderabad Forest Magazine,* (Quarterly), no. II, July, 1942, pp. 5–7.

purpose.[72] Further, in many parts of the State, the area fixed for *bancherai*, being 10 per cent of the total land of the village, was not given to the full. For example, in some villages of Warangal district, it was only 4 to 14 per cent, while it was 10 to 15 per cent in Karimnagar,[73]

Utilization of Forest Products

Utilization was considered as a very important branch of forest economics. Under it the Forest Department had to find out the inherent or intrinsic value of forest products termed major and minor. The government established timber depots and saw mills for supplying industrial raw material, local requirements and agricultural implements in almost all forest divisions. The forest areas of Hyderabad State, being situated in the tropics with different ecological factors within the Krishna valley and the Godavari basin, grew several forest species. Teak was universally adopted as the tree of commercial importance. Shisham and satin were regarded as fancy cabinet and furniture timbers, whilst there were many soft woods of which only bombax was used for match industry and palakordsha and poonka for toys and slate-frames.[74]

For the commercialization of forest products, a special forest officer was recruited in charge of utilization work from 1938 onwards. Information about the demand and supply, the imports and exports, the selling rates in different markets and movement of timber traffic on the Nizam State Railway were collected and coordinated. A great demand of sawn timber was noticed by the Forest Department on account of new designs of buildings. Moreover, the unsoundness of teak logs, due to past-unchecked fires and unregulated selection felling, necessitated the logs to be properly sawn to the sizes, required for the trade. The Forest Department had therefore instituted two saw-mills, close to the big timber forest areas in the Godavari valley, one at Mancherial and the other at Jannaram. They were run by private contractors on payment of suitable wages, under the supervision of the Forest Department.[75]

---

72 Agriculturists' Association, *Immediate Agrarian Problems, Hyderabad State,* Hyderabad, 1949, p. 56.

73 HEH the Nizam's government, Warangal Suba File No. 103 of 1338 Fasli (1929), Record No. 3100 of 1339 Fasli, (1930). Growth of grass and fodder crops in the peasants' own lands which was in vogue already both in Telangana and Marathwada offered the only hope for making good the shortage of common grounds. *Revenue Administration Report, (1914–15),* p. 189.

74 ARFD, (1937–38), p. 26.

75 The government established two more large-scale saw mills in the State, making a total

Railways were the major consumers of the major forest produce like timber of teak, nallamaddi, sal, bijasal, bandaru, tirman, eppa, etc., for its varied requirements. After consultation with the Railway Department, a policy was laid down in 1938, under which the timber requirements of the Railway Department should be first indented upon the Forest Department. This agreement increased the consumption and utilization of local timbers especially bijasal, bandar, nallamaddi and eppa. After testing the timbers at the Secunderabad station in 1938, the Chief Engineer of Railways had sent a list of timbers used on railways to the Forest Department showing species and quantities.[76] Consequently in the year 1939–40, 10,000 sleepers were supplied to the Nizam State Railway, particularly of bijasal and nallamaddi species. Tirman was used for the supply of axe and tool handles by the Railway Department and collieries.[77] With the opening out of Kazipet – Balharsha railway line on account of proximity to Delhi, the Forest Department started timber business transaction with other big centres of consumption such as Delhi. Introduction of state timbers in the foreign and local markets, which were hitherto importing foreign timbers, was also started. In order to supply big and sawn timber requirements in the markets along the Hyderabad – Manmad railway line, a sub-depot was established in 1939 at Nizamabad Divisional Office. One more branch depot was opened at Hyderabad for the sale of sawn material in Hyderabad and Secunderabad markets.

Several non-timber forest products became so important commercially that the Forest Departments at all India level produced a steady flow of research and publications, as well as developing their management in great detail. As a result they turned to be an important source of revenues to the Forest Departments of the British provinces and the native States.[78] One of such commercially major non-timber forest products was bamboo in the forests of the Hyderabad State. Bamboos were found in abundance in the North Eastern districts of the Hyderabad State, i.e., Adilabad, Karimnagar and Warangal. This tree species had a wide variety of traditional uses in the village economy and in money transactions at regional markets. The *Dendrocalamus strictus*, a fibre yielding variety of bamboo, was found to be the

---

of four. Two were located in Adilabad district and one each in Hyderabad city and Nanded district. *Census Report*, Hyderabad State, 1941, p. 179.

76 ARFD, (1937–38), p. 28.

77 'Brief notes on forest utilisation', Conservator of Forests, Eastern Circle, 1939. NAI, Hyderabad Residency Records, Political Branch (1832–1948), Serial No. 926 F – 401 – P. 1948.

78 Ibid.

most useful raw material for paper pulp. The Paper factory started working at Sirpur from April 1942.[79] This was the only important large scale industry that came to be established in Adilabad district, in which bamboo was available plenty in the nearby and surrounding forests, as the principal raw material.[80] It was located at Sirpur on the Delhi – Madras Grand Trunk rail route with an annual capacity of 6,000 tons of paper or 18 tons per day. It had gradually become the largest paper mill in the whole of the southern zone.[81]

In addition to species like bamboo, there were some motley coloured and shining timbers such as the white coloured palakordasha (*Wrightia tinctoria*), the shining satin, the yellow coloured bandar and grey or white coloured mokab, found in fairly large quantities in the forest areas of the Godavari valley. These timbers were utilized for inlaid engravings after smoothening and polishing. In order to advertise articles, manufactured from such fancy and soft woods that created a demand for these, a timber depot and show room were set up in Hyderabad city in 1938.

Cottage Industries

In the state forests there were various kinds of soft woods suitable for match industry, and these soft woods apart from their utility in this industry had no market value whatsoever.[82] The Mahabubia Match Factory which was started in 1928 at Mahabubabad, in Warangal district as a private enterprise, grew rapidly into a large power-driven unit producing the highest number of match boxes in the State and providing employment to a sizable population.[83] This factory also utilized soft woods for manufacturing packing cases and slate

79 *Annual Administrative Report for 1351 Fasli (1941–42),* p. 95.

80 National Archives of India, Hyderabad Residency Records, Political Branch (1832–1948), Serial No. 709, F249 (1) P, 1946.

81 It was spread within a radius of 1,012 sq. miles to Sirpur paper factory and spread over an area of 36,000 acres, with an average of 4.5 tons per acre; *The Golkonda Patrika,* 23.2.1939. For a detailed study on this factory see, L. S. Habib Ahmed, *An Economic Survey of the Sirpur Paper Mills Limited*, unpublished MA (Economics) dissertation, Osmania University, Hyderabad, 1960.

82 There were 25 varieties of soft wood trees found suitable for match industries in the state: Tapary, Thoni, Chidipa, Gargu, Anduku, Punkiki, Buruga, Gumartol, Gummpany, Peddemanu, Peru, Pachimanu, Peddakamba, Poka, Mulu, Thilla, Gunerteel, Burja, Pandiki, Pachichettu, Jirijidi, Mamidi, Thella Poniki, Konda Gogu. Forest Department, Instalment No. 265–1336 Fasli, File No. 1/86, Record No. 1107 / 30 Fasli, (IRH), 1927.

83 Forest Department, Installment No. 265–1336 Fasli, File No. 1/86 Record No. 1107/ 30 F, (IRH), 1927.

frames. It paid a royalty of Rs. 1 per cartload. *Bombax malabaricum* (burgu) was extracted in large quantities to supply the Mahabubia and other match factories. The Forest Department wanted to replace it, as it was feared to be depleted from the forests soon. On the advice of L. Mason, Forest Department recommended all match factories to use *Ailonthus excelsa* which was found a fast growing one, as a substitute for burgu.[84] However, by the year 1943 the scarcity of raw material for match factories was seen.[85] The production of match industries was also hampered by the World War II and later by political disturbances in the State.[86] Lac cultivation and sealing wax manufacturing unit was manufactured at Mamda, in Nirmal division of Adilabad district. It supplied sealing wax to Postal Department, Nizam State Railway and stationery depot of the government.

Amongst the other minor forest produce, ebony leaves, used in the manufacture of beedi, played a very important economic role. Production of ebony leaves brought a good income, as they were not only consumed in the Hyderabad State and served as a means of livelihood for several poor classes, but were also exported to big centres of consumption, especially in Madras Presidency. Charcoal industry flourished in the Medak division due to proximity of forests, connected by a good network of roads and the ever-increasing demand in Hyderabad. The Forest Department attempted to demonstrate manufacturing of briquettes from charcoal dust to the timber contractors.[87] Bee-hiving or apiculture formed a part of Mahabubia Forest School course. Honey was collected and purified for sale in Hyderabad market. The modern hives were purchased for introduction and colonies were also imported for distribution. *Katha* was manufactured from *Acacia sandra* or *Acacia catechu* trees. Its uses were many: dying and tanning in Europe and the Americas, in traditional folk medicine for mitigating fevers and other purposes and all over India it is used in chewing pan (betel leaves). Soft and fancy woods were generally used in toy making. Nirmal was well known for this industry. Medicinal herbs represented the widest variety of species for human use of any category in the Deccan region. Forestry and Botanical Survey documents usually indicated the provenance of each species and its methods of harvesting, preparing and storing.

---

84 'Brief notes on forest utilisation', Conservator of Forests, Eastern Circle, 1939.

85 HEH the Nizam's government, Finance Department, Resource Branch, S.No. 66, File No. 58 of 1355 Fasli (1945–46), pp. 10 and 21–24.

86 *Meezan*, 5.9.1946.

87 ARFD, (1938–39), p. 30.

Thus, the origins of forest management and protection in the Hyderabad State lay in concerns about the depletion of timber of earlier decades by timber traders. The Forest Department was set up after realizing the commercial importance of forests and to check the uncontrolled exploitation by timber traders. The department took over about 12 per cent of forest area for state management. Gradually the Forest Department became a largely commercially oriented department without developing any attitude of working for the development of the people and sensitivity for their needs, except in some fragmented cases. As the timber and other forest products became a source of revenue to the State, the concern for people's needs were complemented by concessions, free-grants, and agrarian extension. The forest revenue in the Hyderabad State was of minor significance when considered individually, but with other departments it collectively occupied a very significant place in the revenue complex of the State.[88] Further, the government had always stated that high priority was given to the environmental needs of the country. Even after taking over of all major forest cover into reserved, protected and open forests, the State could not achieve the target figure of one-third of total geographical area under tree cover. The Nizam's government always gave top priority to conservation in the policy statements, but in practice it could never achieve the goal set by itself.

Talking about the issue of state management of forests as a watershed, the question has to be posed at several levels. In the region of Hyderabad State, no other previous native ruler or State did have a sustained policy of exploitation and regulation of forest tracts. But this was introduced by the Nizam from the second half of the nineteenth century following the British. Introduction of new means of communications and transport, especially the railways, made some changes in the management of natural resources. The creation of a separate department for the management of forest areas was a new phase in the history. At all India level, imperial officials had a wider agenda of intrusion into forest areas than any other previous rulers. The control over forest resources by the State from the second half of the nineteenth century marked the beginning of a new phase in the history, both at all India level as well as in the Hyderabad State. Even before the State became aware and started regulating forest resources, the process of exploitation on

---

88 The revenue from the Forest Department constituted only one per cent of the total revenue of the Hyderabad State. For example, the total revenue of the state in 1920 was Rs. 7.06 crores and the forest revenue was only Rs. 7.14 lakhs; and in 1940 it was Rs. 9.60 crores to Rs. 14.01 lakhs. For a detailed study of the finances of the Hyderabad State see, B. K. Narayan, *Finances and Fiscal Policy,* pp. 67–68.

commercial lines reached high and drew attention of the government. The forest management after the establishment of the Forest Department was one of 'belated and inadequate efforts' to restrain the impact of pressure on forest tracts. Extension of cultivable land, royal/elite hunting and land colonization existed long before the state management of forests.

By this time working plans were designed to do precisely what the phrase suggests – provide a long-term plan for managing forests. Working plans laid down the time between felling cycles and the amount of wood to be removed by the Forest Department. However, many a times the Forest Department itself violated the working plan rules for more production. Economic exigencies rather than ideology often dictated forestry practice. Success was evaluated in terms of the out-turn of 'valuable' species. The strict regime of rotational felling was often violated by the Forest Department that had the duty to implement it. In some ranges, the rotation period was shortened from 30 years to 20 years so as to meet increased demand for timber. Working plans provided for the protection of several tree species that were important sources of revenue. Teak, nallamaddi, bijasal, tendu, etc. were protected for various uses such as making railway sleepers, pit-props, cart axles, furniture, other construction, etc.

The ecological context of state forestry was complex enough for it to be more than a mirror of political and economic issues. 'Species control' was not among the complicated problems in the Hyderabad State. The mixed deciduous forests in the State had a rich variety of trees and plants which were valued by both the Forest Department and the people. But this does not mean that there was the existence of harmony. The selective manipulation of trees and closure of forest tracts for commercial exploitation often led to conflicts between state preferences and rural subsistence users. The working plans, silvicultural operations, conservation methods, management, training and education, exploitation, administrative set-up, acts, rules and regulations, etc. were all nothing but replica of the British Imperial system.

Discontent with the new forest rules and regulations manifested itself in many ways. The option of flight was considered by tribal communities in Khammam district to the Madras Presidency. Another form of protest was non-compliance with imposed regulations. It was evident when peasants and tribals gave misleading information at the time of settlement of rights; 'malicious incendiarism' in the forest areas was one of common discontent. The 'wilful incendiarism' showed the unpopularity of the forest settlement and state management of forest areas. The existing laws did not punish the offenders, owing to lack of evidence and other legal points. A forest guard living in the midst of villagers, did not want to incur their chagrin for the sake of a few law suits in which he was not quite confident to succeed. If he were to succeed he did not entertain much hope of his superior officers rewarding him for what he had been doing.

However, the major lapse at that time was that the State did not make any solid effort to develop local institutions for the people who could then participate in the management. The people also did not develop any affinity towards the forest management meant for meeting their needs as these were controlled either by the Forest Department or the Revenue Department or the local landlords. They continued to use them as open access resources, which, due to increasing pressures, resulted in total degradation or destruction of the resources. However, the state management of forests marked the beginning of a new historical phase in the environmental history of India.

Highlights of the State Management of Forests

Commercialization was the objective behind the management of forest areas in the State. Only marketable timber species received more attention. The rest including all other trees, shrubs and herbs were treated as weeds, no matter what their social, cultural and ethnic values were. This resulted in radical interference with the ecology of forests leading to reduction in the population of all non-timber species, and extinction of vulnerable species.

- Deprivation and marginalization of forest dwellers leading to their migration, undermining of tribal culture and breaking down of mutually reinforcing relationship between forests and forest dwellers
- Deterioration in the quality of habitat of wild animals, birds and insects
- Encroachments on forests became extensive; forest offences increased considerably
- Extensive blanks all over forests due to war felling
- Low priority for investment; little plantation activity

CHAPTER 5

Development of Irrigation and its Influence on Forests

Irrigation is one of the basic ingredients of agricultural activities. In the State of Hyderabad, where agriculture had been under the mercy of monsoon, irrigation acted not only as a stabilizing factor but also as an important production input in any agrarian system. The Hyderabad State lying in the subtropical area was always subjected to scarcity of rainfall and thus a great part of the country was open to severe famine conditions.[1] The earlier rulers, realizing the importance of conservation, control and utilization of the waters flowing down many rivers and numerous streams, constructed many types of water harvesting structures. Tanks were the most popular ones amongst them. In fact tank irrigation had become by far the most predominant in the Hyderabad State, particularly in the Telangana region, where the rivers were not perennial and the country had to depend entirely upon the rainfall for its water. But the monsoon was generally confined to the few months of the year. Though normal rainfall was sufficient to increase the crop production there could be no assurance to the cultivation of both foodgrains and commercial crops without storing water for irrigation.[2]

The heavy sand content in the soils of the Telangana region also promoted the peasant to depend on this source of irrigation. In the trap region like Marathwada, deep and porous soil was not suitable for irrigation. Out of

1 The details of famines in the State will be discussed in the next chapter.

2 *Report of the Royal Commission on Agriculture in India,* Bombay, 1928, p. 324.

nearly 3000 sq. miles of the irrigated area in the State more than 70 per cent was in Telangana. It was called 'the land of rice and tanks' because of its rich soil and well-developed irrigation system. The 18,000 tanks of the State were almost entirely within this area and nearly every suitable site was occupied by a tank.[3] The Warangal district had the most notable examples of the ancient work of irrigation and the lakes of Pakhal, Ramappa and Laknavaram were constructed in the twelfth and thirteenth centuries.[4]

Similarly many other minor water sources of ancient origin existed in the Telangana, Karnataka and Marathwada regions. Infiltration galleries with collecting well and distribution channels were also constructed in the past, examples of which still exist in the Bid district. The other works of importance that deserve mention were the Mir Alam multiple arched dam, Hussain Sagar dam and the infiltration galleries of Aurangabad.[5]

Till the first half of the nineteenth century the maintenance and supervision of the tanks and wells, which were the main source of irrigation, was the responsibility of *Talukdars* (District Collectors).[6] In 1861, Salarjung I took some measures in the direction of irrigation development. He organized an Irrigation Board under the Revenue Department in 1866 as part of his administrative reforms. The vision and talent to organize a Department of Public Works in the year 1868 and the creation of a separate Irrigation Branch in 1869 under the Chief Engineer for Irrigation, showed the most important step in the progress of irrigation in the State. As a result of it, the peasant was to be relieved from the necessity to depend on the rains and monsoons. Salarjung I assigned the duties of maintenance of tanks to Public Works Department in 1867. This department developed gradually, and in 1880 a separate department for irrigation was created under the jurisdiction of the Chief Engineer. The most important step in the Public Works Department during the year 1884 was reorganization of the department which was

---

3 S. P. Raju, Executive Engineer, PWD, *History of the Nizam Sagar Project, Hyderabad–Deccan*, Government Central Press, Hyderabad, 1939, p. 1.

4 K. Balendusekharam, *The Andhra Through the Ages,* Sree Saraswati Publishers, Hyderabad, 1973. p. 44; N. Venkataramanayya and M. Somasekhara, 'The Kakatiyas of Warangal' in G. Yazdani (ed.), *The Early History of the Deccan,* parts VIII, XI, New Delhi, 1982, p. 677.

5 JDB, Gribble, *A History of Deccan,* vol. I, London, 1896, p. 214. The Osmania Graduates' Association, *Development of Hyderabad (The Part Played by Engineers),* Hyderabad, (no date), p. 6.

6 George Palmer, *Report on the Operations of Public Work and Allied Department, 1844–45*. H.H. Nizam's Government, , P.W.D. Press, Hyderabad, p. 6

kept under the supervision of the Chief Engineer.[7] The scarcity and distress of the years 1896 and 1897, that almost pushed the State into a severe famine due to delayed monsoon gave an impetus to the irrigation activities.[8] The threat of repeated failure in rainfall and the attended scarcity and famine resulted in taking initial measures by Mir Mahabub Ali Khan, the Nizam VI, in the direction of irrigational development.[9]

The early programme of the Irrigation Department was to repair, maintain and extend older works. Further, dates and proper statistics of the water resources of the State were collected. But after the visit of Irrigation Commission in 1902–03, the department took up an extensive investigation works of major and minor projects under the guidance of Ali Nawaz Jung, who was aptly considered as the pioneer of modern irrigation in the State. The unprecedented flood of September 1908 that devastated considerable portion of the city of Hyderabad was said to have been instrumental in ushering in what was called as an 'Era of reservoirs' in the history of the State. The immediate sequel to the flood was the institution of an enquiry into its causes and the methods of prevention. This was carried out by M. Visveswaraya and Ali Nawaz Jung Bahadur.[10]

Though some measures were taken up prior to first decade of the twentieth century, the acceleration in the development of irrigation systems began in 1911 with the reign of the seventh Nizam, Mir Osman Ali Khan. He laid the foundation for the major, medium and minor irrigation projects in the

---

7 During this period, Nawab Ali Nawaz Jung Bahudur, Chief Engineer and Sir Mokshagundam Visveswarayya played significant role in designing large scale irrigational projects in the State. K. L. Rao, *Rivers and Their Development,* Waltair, 1964, p. 88.

8 An important event in the history of irrigation in India was the appointment of Indian Irrigation Commission in 1900. When the Indian Irrigation Commission visited Hyderabad in 1901–1902, the state was just recovering from the effects of one of the worst famines on record (1896–97). This was an important event in the history of the development of irrigation in India, as the findings and the recommendations of the commission, which they submitted in 1903 and which were mostly accepted by the government, gave a great impetus to the construction of new irrigation works throughout the country including the Hyderabad State. K. L. Rao, *Rivers and Their Development,* p. 88; K. L. Bhatia, *Development of Irrigation in India,* New Delhi, 1965, p. 83.

9 The Osmania Graduates' Association, *Development of Hyderabad,* pp. 2–3.

10 Recommended preventive measures were: a) enlargement of river channels b) building of storage reservoirs, dams, combining irrigation with flood prevention and for water supply to the city of Hyderabad. As a result two large reservoirs, Osman Sagar and Himayat Sagar were built on Musi and Easi rivers respectively, as flood absorbing reservoirs. M. Visvesharaya, *The Flood of 1908 at Hyderabad: An Account of the Flood, its Causes and Proposed Preventive Measures: A Government Report,* Hyderabad, 1909, pp. 3–12.

State. Tanks and wells were treated as the minor irrigation works. An irrigation project with command area of 2000 acres and above but below 10,000 acres for cultivation was categorized under medium irrigation project. Irrigation was characterized as the major work when acreage exceeded the limit of 10,000 hectares. An irrigation development plan was drawn up on this basis, envisaging long, mid and short term programmes to be completed in one, three and five years respectively.[11]

Irrigation Projects

Minor Irrigation

The minor irrigation facilities such as tanks, puddles and wells formed the backbone of the irrigation system in the State. Tank irrigation was an old established practice and constituted the most important minor irrigation system. Tanks were the part of rural life and culture from the earliest times and they were treated as the common property for all the village people. Other minor irrigation such as wells, were developed by a single family or small groups of men. The indigenous tanks and wells provided water for the otherwise ecologically suited crop. The Nizam's government fully realizing the importance of irrigation had opened a separate Well Sinking Department under an expert engineer.[12] Government recognized the wells as a measure against famine and offered the loans to ryots in the shape of loans called *taccavi*. By 1944 there were 25,451 serviceable and 6,793 unserviceable tanks in Hyderabad State and the number of serviceable and unserviceable wells was being 2,11,529 and 60,025 respectively.[13]

Medium and Major Irrigation

After the establishment of separate Irrigation Department, the State government gave utmost importance to the development of all types of irrigation projects in the Hyderabad State. The Nizam government wanted to utilize the river water for more useful purposes. So major irrigational projects were designed to develop agriculture and combat the occurrence of famines.[14]

---

11 Sadath Ali Khan, *Projects for Prosperity,* Hyderabad, 1951, p. 14.

12 A. I. Quereshi, *The Economic Development of Hyderabad,* vol. 1, Orient Longman Ltd., Madras, 1941, p. 96.

13 Sadath Ali Khan, *Projects for Prosperity,* p. 14.

14 S. P. Raju, *History of the Nizamsagar Project,* p. 1. The major irrigational projects like the Nagarjuna Sagar dam which is the world's largest masonry gravity dam, was under contemplation since 1903 by the Hyderabad and Madras governments. It was finally started after independence and completed by 1979. The project is located at about 2 kms from Nandikonda village in Nalgonda district.

By the year 1928, the Irrigation Department repaired and constructed minor and moderate sized irrigation works and brought a total area of about 8 lakhs acres under irrigation. But irrigation in the Hyderabad State was mostly confined to Telangana as 80 per cent of the area under irrigation lay in this tract.[15] From this year onwards, intensive drive was made towards the construction of major and medium sized reservoir projects. The important amongst these either completed or under completion, during the period of the present study, were Nizamsagar, Tungabhadra, Dindi, Palair, Wyra, Rooty, Pendlipakhla, Chandersagar, Baithpalli, Kadam, Pocharam, Koilsagar and Singabhupalem. In addition to the above mentioned projects there were some other minor projects in the State either completed or launched in the late 1940s like Akher, Jowlinala, Halia, Bendsura, Kamli and Talwar, Khasapur, Lankasagar and Edalvagu, Chintal Cheru, Maner, Kadam, Yakinour and Chagvan, etc.[16]

These were by far the most important form of irrigation facilities in the Hyderabad State that were encouraged by the government. Almost all the irrigational projects were located in the Telangana region of the State. Irrigation facilities had developed mainly in Warangal, Karimnagar, Nizamabad, Nalgonda, Mahabubnagar, Medak and Adilabad districts. The Kuntas or the percolation tanks also formed the largest, but served as indirect source of irrigation in the region, as they kept the water table high in the vicinity.

Development of Cultivated Area

In the first decade of the twentieth century, cultivable area increased from 204.07 lakh acres to 227.95 lakh acres. The increase in cultivated area was 11.60 lakh acres between 1912–13 and 1920–21. This increase was due to reclamation of cultivable land, diversion from forest area or grasslands, etc. Dry land cultivation increased during 10 years ending 1912 by about 26.40 lakh acres, and in the next decade it increased by 8.04 lakh acres. The increase in wet cultivation during the same period was 1.97 lakh acres and 0.48 lakh acres respectively. The percentage of wetland to the total cultivated area remained almost stationary during the first two decades of the twentieth century.

Regarding irrigated area, in 1902–03, according to the estimates of the PWD, 10.43 lakh acres of land were under irrigation. This figure increased

---

15 A. I. Quereshi, *The Economic Development,* pp. 88–92.

16 The Osmania Graduates' Association, *Development of Hyderabad,* pp. 6–9. Abdul Quayyum Khan, *Irrigation Hyderabad State,* in Hyderabad Government bulletin in Economic Affairs, Vol. I, No. 9, (December, 1948), p. 27.

to 12.08 lakh acres in 1910–11, but went down to 7.88 lakh acres by 1921–22. In 1902–03, government tanks and canals irrigated 7.9 lakh acres, and 2.51 lakh acres were irrigated by wells. Irrigated area increased further to 12.78 lakh acres by 1930–31 and 14.62 lakh acres by 1940–41 and went down to 13.77 lakh acres by 1944–45. Between the years 1947–51, on an average it went up to 14.05 lakh acres. During the last 30 years, irrigated area in the State rose from 10.50 lakh acres to 14.05 lakh acres. During 1940s the districts of Nizamabad, Medak, Mahabubnagar, Nalgonda, Karimnagar and Warangal had over 70 per cent of the total irrigated area in the State. The area under cultivation was 237 lakhs of acres in 1948 as compared with 96 lakhs of acres in 1870s. The following table 1 gives quinquennial average information on cultivated and irrigated areas in the State.

Table 1. Proportion of irrigated area to total cultivated area (Quinquennial Averages)

Year	Total cultivated area (acres in lakhs)	Irrigated area (acres in lakhs)	Percentage of irrigated area to area cultivated
1922–26	266.96	10.48	3.92
1927–31	277.31	10.85	3.91
1932–36	291.41	12.65	4.34
1937–41	275.96	10.80	3.91
1942–46	264.91	13.48	5.08
1947–51	237.49	14.05	5.91

Source: Administrative Reports and Statistical Year Books, relevant years.

Of the total cultivated area, about 95 per cent was rain fed or dry crops. The irrigated or wet area was nearly 5 per cent, largely in Telangana region where irrigation sources were in abundance. Though wells and river projects were there, other sources of irrigation such as tanks, both small and large, surpassed them in the extent of their contribution to land irrigation. The number of tanks and their state of repair showed a fair contribution of the prosperity of the State. But, the restoration and maintenance of these tanks left were largely neglected. Consequently, the ayacut under them got greatly diminished, and the State found itself in the grip of a formidable food crisis, spelling disaster for the people.

Cropping Pattern

Irrigation and rainfall and to some extent prices of crops were the most responsible factors for the cropping pattern in the State. Particularly, the influence of irrigation and rainfall on the cropping pattern was crucial in the State. The prices, however, did not cause any influence. Irrigation provided a congenial environment to the farmers to grow high yielding crops. Farmers continued to grow subsistence crops in the absence of markets, and transport facilities.

The irrigational development in the State attracted people from coastal and Rayalaseema districts. Rayalaseema settlers introduced the cultivation of the most important cash crops in Telangana in the shape of groundnut.[17] Those from Krishna and Guntur districts were responsible for the starting of cultivation of tobacco, chillies and sugar cane.[18] Non-food crops formed roughly one third of the net sown area, as they brought a great transformation in agricultural economy of the State. The acreage under food crops declined from 160.76 lakh acres in 1922–26 to 158.15 lakh acres due to the shift towards the commercial crops. In fact, the area under commercial crops like cotton, groundnut, castor and chillies increased from the end of the nineteenth century. In the beginning of the twentieth century, and more particularly after the First World War, the area under commercial crops increased.

The most important food crops sown in the State were jawar, bajra, wheat, rice and other cereals, minor millets and pulses. The average for the 5 years during 1936–40 was 70.15 per cent of the total area under cultivation. Acreage under rice was about 3.6 per cent of cultivated area of the State. Jawar covered the highest percentage with 30 to 35 per cent of cultivated area. On the other hand, area under wheat was around 4 per cent which was mainly grown in Marathwada tracts. Over 60 per cent of the wheat sown area was in Aurangabad, Parbhani and Nanded districts. Bajra was the next important crop. The area of bajra cultivation decreased from 28 lakh acres in 1925–26 to 16 lakh acres in 1939–40. The total area under pulses was around 30 per cent. The most important among the non-food crops as well as commercial crops was cotton which was sown over 12.57 per cent of the total gross area. Next commercial crop in the series of importance was groundnut which was sown over 6.7 per cent of the total gross area sown.[19] The net sown area under food crops showed a declining trend as the years progressed and the fall was paradoxically more marked since 1940s. This was probably due to the peasants being drawn towards raising rich-yielding money crops. To stop this undue advancement of cash crops, the 'Grow

---

[17] B. N. Chaturvedi and K. Surender Reddy, 'Irrigation in Telangana sources and problems' in *Deccan Geographer,* Hyderabad, January 1964, p. 171.

[18] *Hyderabad Information,* vol. II, No. 9, 1351 Fasli, June 1942; also see N. G. Ranga, *Economic Organisation of Indian Villages,* vol. II, Bombay, 1929.

[19] Department of Agriculture, HEH the Nizam's government, File No. 32–8, S. No. 97, 1948.

More Food' campaign was launched by the government in 1942. Advances were made as *taccavi* in the shape of improved seeds, manure and implements were granted to grow more food crops.[20] In spite of the launching of this campaign to bring more acreage under rice and other food grains production, the acreage under non-food crops increased in leaps and bounds.[21]

Migration and Growth of Population

There was migration of cultivators from coastal districts during the 1921–41 and the Nizam government also induced experienced cultivators from coastal Andhra to introduce new methods and techniques of cultivation in the State, under colonization policy. Most of these cultivators settled down in Nizamabad, Khammam and Warangal districts, particularly in the areas near by big irrigation projects. They were known for their innovative and intensive cultivation.[22] In order to irrigate fresh wet lands it was found necessary to encourage cultivation from other places in the State to settle down in the ayacut areas of irrigation projects like the Nizamsagar. To achieve the end a colonization scheme was drawn up and a special officer was appointed. Nearly 3,000 families migrated and settled down in their new homes to irrigate the lands that were allotted to them by the government.[23] The soldiers, who rendered the service to the Nizam, were also allotted the lands in different areas under the irrigational projects. There was a criticism that they showed more interest in trying to reap the profits out of the allotments rather than put the lands to fuller and proper utilization.

Effect/Pressure on Forests

Among the effects of the development of irrigational facilities, the most significant one was large-scale changes in land-use pattern. The area under cultivation extended considerably after the construction of irrigational projects and other irrigational sources, often at the expense of forests and grazing lands. In this process many state forest-lands were excised for cultivation. By 1900 forests had shrunk considerably; and by the first four decades of the twentieth century all the forest areas were under commercial exploitation. But no clear planning concept lay behind the clearance of

---

20 Ibid.

21 *Hyderabad Information,* vol. IV, no. 10, (Shahrewar, 1353 Fasli), July, 1944, p. 10.

22 There was a massive migration into Telangana from Coastal Andhra, see N. G. Ranga, *Economic Condition of Indian Villages,* Vol. II, Bombay, 1929.

23 Sadath Ali Khan, *Projects for Prosperity,* p. 37.

forests. The revenue officials offered forest-lands for extending cultivable area at their own discretion, even sometimes without consulting the forest authorities. The peasants were encouraged to get more land under cultivation in addition to the increasing demographic pressure. Foundation of new forest villages provided further impetus. Consequently, forest area in districts like Nizamabad decreased faster than that of the average of the Hyderabad State.

The following table 2 shows in a comparative perspective the development of cultivated area in the State and the corresponding negative trend in the growth of forest area which began to decline gradually. The rise in the cover under forest area during 1937–38 was due to inclusion of wastelands, but not because of increase in the forest area. On the whole it declined from 79.08 lakh acres to 61.70 lakh acres from the beginning to the middle of the twentieth century.

Table 2. Progress in cultivated area and forest area between 1893–1946 (acres in lakhs) (State total area 5,29,26,720 acres or 82,698 sq. miles)

Years	Forest area (Percentage to total area)	Cultivated area (Percentage to total area)
1893–94	N.A.	161.45 (30.50)
1902–03	N.A.	176.78 (33.40)
1912–13	79.08 (14.94)	205.16 (38.76)
1917–18	60.98 (11.52)	212.02 (40.05)
1922–23	59.91 (11.31)	214.23 (40.47)
1927–28	60.99 (11.52)	209.62 (39.60)
1933–34	62.84 (11.87)	293.58 (55.46)
1937–38	63.99 (12.09)	263.71 (49.82)
1944–45	61.70 (11.65)	264.75 (50.02)

Source: Compiled from *General Administration Department Reports* and *Statistical Year Books* of relevant years.

The dynamics of arable expansion differed according to the irrigational facilities, availability of land, nature of soil and terrain and other factors. The above statistics show that the forest area decreased from 79 lakh acres in 1912 to 61 lakh acres in 1945, while the cultivated are increased from 205 lakh acres to 264 lakh acres. The percentage under forest cover declined from 14.94 to 11.65, whereas the cultivated area increased from 30.50 per cent to 50.02 per cent. The agrarian expansion not only took place in forest areas but also in other wastelands such as banjars, scrublands, gairans and porambokes. It is noticeable that cultivation always expanded from water sources to dry land areas. The irrigational reservoirs watered the command areas and some lands were double cropped too. In the command areas of the dams, cultivation moved onto virgin lands which had good fertility, whether it being forest or banjar lands. Existence of large tracts of cultivable land around villages made the expansion possible. Cultivation was normally done around the villages and settlements and it expanded in all directions. Expansion also took place onto the dry areas with poor quality soil, where coarse grains were grown. However, the area under coarse grains in the State was much larger than under wet cultivation (95 per cent of the total cultivated area was under dry cultivation). This indicates that expansion was more on poor quality dry lands.

Revenue Department vs. Forest Department

There was little evidence of conflict between agriculture and forests in pre-colonial times, as the land's resources were more than adequate for a sparse population. But, Nizam's government policy, being influenced by the British revenue policies, started opening new lands for agriculture, first as a source of increased revenue, and later as a recognition of the needs of a growing population. Till the middle of the nineteenth century, all eastern Telangana districts were heavily of forested and lightly populated. No one as yet foresaw any danger to the land's resources if revenue, commerce, and food production were all encouraged.

In the Hyderabad State Revenue Department had equal powers with the Forest Department to take decisions regarding the open forests. However, it did not have good establishment to supervise them. All they did was to issue free passes for as much timber as the ryots chose to ask for and to auction forest produce generally for much below its value, as they were not fully acquainted with its uses and value. Forest Department thus alleged that the Revenue Department's development policy under the schemes like *Ijara* and colonization scheme had been largely responsible for deforestation in the late nineteenth and the beginning of the twentieth century.[24]

S. H. Hussain, Divisional Forest Officer, Medak division, whose opinions reflected the Forest Department's stand, lamented that the capitalist and influential classes sought various means to grab more land. They found, according to him, forested areas to satisfy their greed for the accumulation of land under the name of *nao-laoni* and 'extension of agriculture.'[25] In this process thousands of acres of forest-land was given away on *patta* with the result that in majority of cases the *pattadars* cleared the forest growth and left

---

24 *ARFD,* (1900–1901), pp. 2–3. And S. H. Hussain, *Forest Policy for the Dominions and its Necessity*, p. 1.

25 S. H. Hussain, *Forest Policy for the Dominions and its Necessity*, p. 1.

the land uncultivated. Later the colonization scheme and Development Department accelerated the process of clearance of forest areas for agrarian extension. Various reasons were brought forward to get valuable forest areas on *patta* and the Forest Department complained that the government almost always sided with the applicants. In later years dairy farms, to which the Forest Department called as the new species of land grabbers, took the colonisation scheme and many forests were given away in spite of strong objection from the Forest Department. The department accused that many *ijaradars* and *pattadars* were not interested in the development of land for cultivation but interested in the valuable forest growth.[26]

The view of the Forest Department was different from the Revenue Department with regard to agrarian extension. It reported in 1893–94 that 'cultivation has unnecessarily been permitted by the revenue officers in the forests, and under the cloak of clearing land for cultivation, much valuable timber has been destroyed.'[27] Sometimes it granted some forest lands for cultivation after scrutiny of the applications. The Forest Department boasted its activities, as 'it was quite willing to part with any area whenever and wherever necessary in the legitimate interests of the country and the people'. It showed an example of 400 acres from the Royenpally Reserve given for cultivation.[28] Greater expansion of cultivation took place between the years 1913 and 1927 mainly in the Telangana region. The reason behind it was the existence and construction of large-scale irrigational projects in the area. Ablout 2,22,646 acres of forest-land brought under cultivation in Marathwada region during this period, whereas 9,77,861 acres was brought under the cultivation in Telangana. In 1913, only 1 per cent of assessed area remained for occupation in the former region as against 20.7 per cent in the latter.[29]

However, the Revenue Department argued differently. It said that the prevalence of thick forests, interspersed between cultivated lands and their non-demarcation and clearance, had not only restricted the irrigable area, but on the other hand, immensely contributed for the prevalence of acute malarial conditions. This in turn, resulted in sparseness of population and scarcity of agricultural labour.[30] These conditions existed in varying degrees,

---

26 HEH the Nizam's government, Warangal Suba File No. 1 of 1357 Fasli (1948) and R. No. 2064 of 1357 Fasli.

27 *ARFD,* (1893–94), p. 69.

28 *ARFD,* (1937–38), p. 3.

29 *ARFD,* (1937–38), p. 3. HEH the Nizam's government, Agriculture Department, S. No. 2 and File No. 56 of 1354 Fasli (1945).

30 The incidence of malaria was so acute in these forest-ridden ayacuts that peasants and labourers came from the adjoining tracts to take up cultivation work in these areas only

in the ayacut areas of Ramappa, Pakhal, Laknavaram and Bhimghanpur. The first *talukdar* (District Collector) of Warangal expressed that if the limits of government land and hunting preserve (*shikargah*) were settled and demarcated, hundreds of acres could be brought under cultivation. According to the Revenue Department thousands of acres of irrigable land thus left uncleared and unutilized in dense jungles and this adversely affected the scope for increased agricultural production.[31] The Revenue Department wanted to examine all degraded forest areas to get it free from the control of Forest Department. If the areas had thick forest cover and they were favourably situated they would be retained as permanent resources. Otherwise the suitable lands would be given away for the agricultural extension. It advised that the sale of forest produce to peasants would bring in revenues to the Forest Department, while the protection of trees would ensure a steady income/supply of inputs for agriculture in future.

Thus, in the state forests there were many areas that were suitable for cultivation and had little potential for forestry. Forest officials had discretionary powers, particularly in reserved and protected forests. They could choose which lands would be opened up for cultivation. Even they tried to retain large tracts of forest-land under their control, and justified this on grounds of preservation of climatic stability. The Forest Department was considered as the most suitable government agency for the management of all government wastelands. Sometimes the department even merged the cultivated lands that had been cultivated for such a long period of 24 to 44 years while reserving the forest-lands into the state forests.[32] Even though alternate lands were shown and compensation was given in some cases, these

---

during tabi season and that too, when they failed to get work in their own and other villages for want of water supply for irrigation. They left these villages immediately after the harvest was over and abi crop could not be raised in these areas even though there was sufficient water in the reservoirs for taking it up. HEH the Nizam's Government, Agricultural Department, S. No. 21 and File No. 56 of 1354 Fasli (1945), p. 4. However, the study of this aspect forms a different story. For a study on this subject in Northern India see, Elizabeth Whitecombe, 'The Environmental Costs of Irrigation in British India: Waterlogging, Salinity, Malaria', in Arnold and Guha, (eds.), *Nature, Culture, Imperialism*. In her pioneering study she had shown how the orientation of agricultural production towards the export market and the construction of large irrigation works which increased soil salinity, water logging and spread of malaria in the Doab region.

31 HEH the Nizam's government, Agriculture Department, S. No. 2 and File No. 56 of 1354 Fasli (1945).

32 Warangal Suba File No. 8 of 1350 Fasli (1941) and Record No. 1477 of 1350 Fasli (1941).

did not satisfy ryots, as they had to work for the development of these new lands with different soil types and fertility.

The process of exclusion of lands from government forests for agriculture was complementary to the re-evaluation of lands in terms of their capacity to generate forest produce. In the reserved forests the Forest Department had full control over them and in the protected forests it had somewhat less control and in the open forests with even few powers. The last types were considered suitable for agrarian extension and for the supply of forest produce to peasants. Forest produce, like fodder and wood was vital for peasants. Some crops like sugarcane required forest props for its support. In the Nizamabad district this cultivation affected the management of forests, as it was heavily dependent for its support on forest props.[33] However, after the 1930s the introduction of new varieties made sugarcane stem sturdier that required no support of forest props.[34]

The vague concept of reserving lands was arisen, as the definition of forest was not clear to the common people of the State. The term could be applied to land even totally devoid of trees and which was suitable for cultivation. The state forest was thus a juridical category but not a descriptive one. Many areas were annexed because there was no one whom the government wished to recognize as the proprietor.[35] The creation of state forests might not be so much for purposes of forestry, as for the alienation of property rights to land. The Forest Acts clearly stated that all wastelands belonged to the State and many tracts were annexed because there was no one to claim the lands as the owner. Gradually the Forest Department got a good say in government decisions regarding allocation of forest-lands for cultivation. The control of cultivable/arable lands falling under the category of reserved or protected forests made the Forest Department an important

---

33 S. P. Raju, *History of the Nizamsagar project,* p. 15.

34 However, revenue and forest officers were divided on the question of who should be granted rights to the access of forest products for agricultural needs, free of charge. Though the forest policy recognized importance of forest products for agricultural and local requirements, foresters were unwilling to accommodate these demands on a large scale.

35 Iyengar recommended in his investigation conducted during 1931–32 that large parts of the scrub forest area could be profitable given away for cultivation. He proposed to establish agricultural experimental farms and demonstration centres which would increase the agricultural yield. But for any experimentation, demonstration or propaganda scrub forest area must be first cleared. The Forest Department would then be able to concentrate upon the real net forest area as such around and along hill range and away from agricultural hamlets. S. Kesaya Iyengar, *Economic Investigations,* vol. III, Warangal district, pp. 4–7 and 22, vol. V, p. 22.

decision maker in the agrarian extension. In the reserved forests the Forest Department had de jure control that the Revenue Department hoped to open up for cultivation. However, in the case of open forests the Revenue Department had also equal control to alienate the land for cultivation. For the *ijara* and colonization schemes many forests were cleared to extend cultivation, but the forest officials claimed that forest wealth should not be irretrietavably damaged in the process. Foresters like Hussain who were vehemently opposed to the idea of agrarian extension in the state forest areas, asked the government that the Forest Department should decide when and where forests could be excised to make way for cultivated land.

The expansion of cultivation was bound to work to the advantage of revenue officers in the long term. However, the Forest Department was strict in the defence of its powers over forested lands. This department thought that the Forest Department failed to understand that forest conservancy must always make way for the agrarian extension. However, the Forest Department had supreme power over the selection of lands to be excluded from the reserved and protected forests while having supervisory powers over the open forests.

Ecological Change

The first signs of climatic change were studied during the 1930s. Contemporary observers like geologist Captain Leonard Munn worked extensively in the Hyderabad State in this regard. He recognized that the failure of rainfall and increase in temperature could be attributed to deforestation in the State. The general rise in temperature and the velocity of winds corresponded with the absence of rainfall. Occurrence of a series of famines in the second half of the nineteenth century, according to him, with more intensity than earlier ones had the cumulative effect of natural and man-made causes. All these famines were more than a natural swing of climatic mood. The severe famine of 1876–78, which had such catastrophic consequences, might have resulted due to excessive deforestation. After more than five decades of haphazard felling the ecological balance had been lost and natural regenerative process no longer functioned mostly in the dry areas of Karnataka region and partly in the Telangana district.[36]

The remarkable thing about the Hyderabad State was that the geologists were obviously aware of the nexus among the area of forest, expansion of

[36] Captain Leonard Munn, 'Geology of the underground water resources of the Hyderabad State and notes on well sinking', *The Journal of Hyderabad Geological Survey: Water-Supply Paper No. 1,* vol. II, part: 2, pp. 4–5.

cultivated land and the decrease in rainfall. Forests had a great influence on the climate and the water resources of any country, especially in the tropics. The hot winds caused considerable soil erosion. Worse, they loosened the surface of the now unprotected soil bare of any vegetation, which was then washed away with the onset of monsoon rains. Canal erosion took place in the headwaters of streams and rivers, finally drying up the springs. During the 1930s, the subsoil, which was crucial for water storage, was found in the dry districts of the state like the Raichur district. Captain Munn, while recognizing the importance of the forests in the State noted in 1934 that:

> The destruction of forests, and especially the entire removal of all trees and vegetation over large areas in the State, with no attempt or scheme for replantation, has already had a very serious effect on the underground water resources. Early steps should be taken to give cultivators every inducement to remedy this error, and tree planting on a big scale in famine areas, especially along lines parallel to the contours, is most strongly recommended. This will not only assist the underground water resources, but act as a part prevention to the serious soil erosion, which is occurring over large areas.[37]

The effect of the deforestation, probably combined with a gradual, nearly imperceptible, diminution of the rainfall over long periods, accounted for the shrinkage of the water table in the south-western, western and central portions of the Hyderabad State. Captain Munn showed the evidence of destruction of vegetation, except those trees protected by the government, in the Karimnagar district, east of the railway line, and south of the Godavary where he lived for 5 years. That area was opened up for cultivation in the beginning of the twentieth century. Local inhabitants, about 60 years of age, told him that as young men, villagers would only travel in parties to Peddapalli owing to the number of tigers and wild beasts in the forest. By the 1930s, except for a few tamarind, mohur, palmyra and toddy trees, protected by the State, the area became treeless, and villagers started delving for roots to obtain fuel. Similar conditions were witnessed in the other districts like Raichur due to the effect of entire destruction of vegetation. This caused a serious diminution of the water table, combined with very serious erosion of the black cotton soil upon the slopes.[38]

---

37 Ibid., p. 5.

38 The diminution of the rainfall had naturally a great effect on the level of the water table. The note published in the Annual Report of the Well Sinking Department for 1340 Fasli (1930–31), illustrated with rainfall graph, map and sections, showed many instances in the western half of the Raichur district where the water table had shrunk 10 feet and over, during the second decade of the twentieth century. Ibid., p. 32.

Munn feared that in certain areas the sand would separate from the binding soil and start drifting with the winds, forming shifting sand dunes in Adilabad district. Instances of this were already visible in forest clearings. Continuous pecking of the surface by herds of goats enhanced these effects further. An early effect of the destruction of vegetation was the development of gullies, or wash ways on hillsides and plains; the former being very noticeable along the scarps of the Deccan trap plateau in the Marathwada region, whereas the latter on all black cotton-soil slopes. These gullies were the active agents for the annual removal of tons of soil from the affected areas.[39] On top of the various types of erosion, there were the effects of ever more intensive farming. After deforestation the exposed soil in the plains rapidly became saliferous. The forest offered sufficient protection against this, but because of the increase in water evaporation salt moved up to the surface and was deposited in the topsoil, leading to a reduction in the soil's capacity to retain water.

The silting up of reservoirs was observed immediately after their construction. In the course of the monsoon mud and the plants filled up the water bodies. During the dry season that followed, most of the water evaporated and sand deposits increased. Thus leftover soil and the remains of the plants and harvests were washed into silt and the whole was compounded by the increase in sand and mud deposits.

On the other hand, high temperature during the summer caused some tropical diseases like malaria. As the water in the newly constructed reservoirs, tanks and kuntas received more sunshine than before, they turned into breeding grounds for mosquitoes where the larvae could develop. Malaria increased drastically in the course of 1920s in the ayacut areas.[40] After the famines and droughts of the second half of the nineteenth century there were renewed epidemics. The diseases like cholera (locally called *gattara*) were, of course, not a result of the deforestation of forested areas in the State as well as India. But epidemics like it indicated a dramatic change in the

39 Ibid., pp. 3–5. However, 80 years is too short a period to come to any definite conclusions. But, history pointed to a great total shrinkage of the water table, which occurred in the last 200 years in the southwestern, western and central portions of the Hyderabad State.

40 HEH the Nizam's government, Agricultural Department, S. No. 21, File No. 56, of 1354 Fasli (1945), p. 4. See Elizabeth Whitecombe's works on this trend in North India, 'The environmental costs of irrigation in British India: Waterlogging, salinity, Malaria', in Ramachandra Guha and David Arnold (ed.), *Nature Culture Imperialism;* and *Agrarian Conditions in Northern India: The United Provinces under British Rule, 1860–1900,* Berkeley, 1971.

ecological conditions of the Indian subcontinent at the beginning of the twentieth century.

Thus, as discussed in previous section, the agricultural frontier was not closed in the Hyderabad State and expansion took place till 1948. Within the State, the area of cultivated land expanded from 161.45 lakh acres to 265 lakh acres between 1890 and 1945. There was a continuous increase in the cultivated area from the last decades of the nineteenth century, but significant expansion took place after 1920 due to construction of various irrigational projects. However, considerable areas of forest were lost during the same period from 80 lakh acres to 61.70 lakh acres in the process of arable expansion. The main reason for agrarian expansion in the State was to cope with the recurrent famines and to feed the growing population (which increased from 98 lakhs in 1881 to 186 lakhs in 1951). Naturally, population growth increased the demand for more food, which in turn drove wasteland and forest clearance to bring more land for agriculture. When population growth increased the demand for food in a technologically limited culture, the natural response was expansion of the cultivated area. Intensive agriculture and spatial expansion were the only options to cope with the population pressure. The increase in arable land area was not limited to the subsistence agricultural sector. Agricultural clearance for commercial crops such as cotton, sugarcane, chillies, etc. in the command areas displaced vast tracts of forest during the period under study. Government with the schemes such as *ijara* and colonization schemes encouraged bringing fresh lands under cultivation. Forest-lands were the most favourable tracts for the expansion of arable land. Though the forest officials throughout the period competed for control of the forest resources against the revenue officials, they could not stop arable extension in the forest areas.

CHAPTER 6

Famines vs. Forests

It is a well-known fact that India from the eighteenth century onwards experienced a series of famines recurring at very short intervals bringing immeasurable sufferings and mortality. Different regions in the country were affected which brought disorder into the lives of millions of people. However, analyses of these famines by historians have rarely included the study of forest zones as the dependent source of sustenance during times of famine. It was overlooked that how forests protected people from starvation and death and how certain communities averted crisis and withstood drought due to their access to natural resources in forests. With the exception of one or two of these types of studies, many works on famines have neglected the impact of state forestry on the traditional methods of survival during scarcity, drought and famines.[1] The present chapter makes an attempt to examine the history and nature of famines and scarcity in the Hyderabad State. In addition to this it endeavours to find to what extent people depended on forests during the times of crisis and the state response to the demands of the people during the famines.

---

1 Scholars like Vinita Damodaran in a significant study on the famine of 1897 in Chota Nagpur region has shown how local/tribal communities had depended on forest resources for survival prior to the colonial forestry closed access to the forests and how State forestry destroyed the traditional economy and brought dislocation into sustainable style of living which made these communities much more vulnerable to famine and disease. See, 'Famine in a Forest Tract: Ecological Change and the Causes of the 1897 Famine in Chota Nagpur, Northern India', in R. Grove, V. Damodaran and S. Sangwan (eds.), *Nature and the Orient,* pp. 853–890 and also see her 'Gender, Forests and Famine in 19th-Century Chotanagpur, India', *Indian Journal of Gender Studies,* vol. 9, no. 2, 133–163 (2002).

The Hyderabad State, in common with other parts of colonial India, was subject to periodical visitations of famine of a more or less severe character. The State had been witnessing famines since pre-colonial times onwards, which was a common occurrence, mostly due to the delay of monsoon rains. The accounts of famines in the Hyderabad State exist from the seventeenth century. Mulla Kamal's *Shah Nama*, Khafikhan's *Muntakhab-Ul-Labab*, the *Alamgir Nama*, *Hadiqat-ul-Alam*, etc. contain information pertaining to famines in the Deccan region during the seventeenth and eighteenth centuries. Famine reports show that in Hyderabad State drought, war and excess of rain acted as the causing factors behind severe disasters and famine situations. Causes of famines and their respective years in the Hyderabad State are given below:[2]

From Drought:	1629, 1659, 1685, 1713, 1787, 1790, 1802–4, 1813, 1819, 1846, 1854–5, 1862, 1866-67, 1871, 1876–78, 1897–1900, 1906, 1912–13, 1920–21, 1940
From War:	1631, 1650, 1682–83
From Exccss rain:	1676–77, 1702, 1825, 1918–19

The above data shows that a majority of famines in the Hyderabad State were due to drought followed by war and excess of rains. The reports on the famines in the State indicate that no particular part of the State can be said to be immune from periodical visitations of scarcity and famines.[3] The western portion of the State comprising districts like Aurangabad, Bid, Osmanabad, Gulbarga and Raichur were, however, more susceptible to crop failures particularly of *kharif* crops than rest of the State. This part of the State had been recognized as a 'Famine Zone'.[4] These districts witnessed a periodical

2 Maulavi Syed Mahadi Ali, *Report on the History of the Famine in His Highness the Nizam's Dominions in 1876–77, 1877–78,* (Hereafter RHF), Exchange Press, Bombay, 1879. HEH the Nizam's Government, Revenue Department, Hyderabad *Scarcity and Famine Manual,* Government Central Press, Hyderabad, 1938; *Karimnagar District Gazetteer,* (1972), pp. 102–3 and A. Loveday, *The History and Economics of Indian Famine,* G. Bell and Sons Limited, London, 1914, (Reprint, Usha Publications, New Delhi, 1985) pp. 135–8.

3 S. H. Bilgrami and C. Willmott, *Historical and Descriptive Sketch of the Nizam's Dominions,* vol. II, Bombay, 1884, pp. 248–50. V.K. Bawa, *The Nizam between Mughals and British, Hyderabad under Salarjung I,* S. Chand and Co., New Delhi, p. 63.

4 The famine zone seems to have been most affected even prior to nineteenth century of which the *Doi-Bana* or the skull famine of 1792 was widely mentioned; HEH the Nizam's Government, Revenue Department, *Hyderabad Scarcity and Famine Manual,* (hereafter HSFM) Government Central Press, Hyderabad, 1938, p. i; *Imperial Gazetteer of India, Hyderabad State,* 1909, pp. 48–49.

failure of crops leading to famine situations. Raichur and Gulbarga were the districts that were generally the first to suffer from a failure of rains, and were more liable to famine than any other part of the State. In many years the failure of rainfall in this 'zone' had led to shortage of food grains, fodder and drinking water resulting in the evacuation of whole villages. However, the systematic records of famines, their course and relief administration in the Hyderabad State were available from the great famine of 1876–78.

The Great Famine of 1876–78

This famine had been described as the most grievous calamity of its kind that the country had experienced since the beginning of the nineteenth century. The distress started with the failure of monsoon in 1876 over a large part of South India. It included large part of the Hyderabad State, half of the Madras Presidency, whole of Mysore and Deccan districts of Bombay Presidency affecting an area of about 200,000 sq. miles and harming a population of around 36 millions.[5] The places in the Hyderabad State actually affected by famine were Lingsugur, Raichur, Shorapur, Gulbarga, Bid and Naldurg districts comprising a vast area of 30,295 sq. miles.[6] In the districts of Nagarkurnool and Nalgonda there was no famine, but the distress caused by scarcity was severe. Though this calamity affected the southern and western parts of the State, scarcity and distress prevailed throughout the territory. In fact, the whole State suffered from the famine, in the shape of high prices of foodgrains, and the influx of famine-stricken people to less affected regions from affected regions. The rise of prices was attributed to hoarding by private traders in addition to low yields due to failure of rains. The average rainfall was 34 inches in the State and during the famine period it was less than half of the average, which later accounted to 15 inches. Relief works were started in October 1877 and were finally closed in November 1877. Salarjung I laid down principles in his minutes of October 1877, which formed the basis of all relief measures adopted. The instructions given in the minutes were briefly as follows:

1. Price of grain was not to be interfered with.
2. Works to provide employment for peasants in need should be opened wherever the necessity was proved to exist.
3. In the districts where crops had not failed, ordinary work should cease and

[5] Government of India, *Famine Commission,* Report, Part I, (1880), pp. 16–17, quoted in A. Loveday, *The History and Economics of Indian Famines,* p. 56.

[6] RHF (1876–77, 1877–78), p. 43.

4. District authorities were to submit weekly reports on all needful points, so that prompt action might be taken where advisable.[7]

Relief measures were adopted by the government during the famine in three ways:

1. by the construction of public works
2. by starting special relief works and
3. by opening poor houses.

A Central Famine Relief Committee was appointed at the commencement of the famine to suggest relief measures with Maulavi Mahdi Ali as the Famine Commissioner. He submitted the famine report in 1879. Notwithstanding all the relif measures taken by the government the death toll could not be avoided. The *Report on the History of the Famine in the Nizam's Dominions in 1876–77, 1877–78* shows the lamentable fact that 71,658 persons died and the mortality in cattle figured at no less than 432,926. However, the actual mortality was probably much higher than the official estimates.

The Great Famine of 1899–1900

This famine was caused by the failure of the southwest monsoon in 1899. In that year the State received only 15.49 inches of rainfall on average, less than half the usual quantity of 34 inches.[8] The deficiency extended more or less to every district. In a total of 71 *talukas,* out of a total of 112, the rainfall was less than half of the average, while in the most favourable instance it was 18

---

7 All the instructions in this minute were enunciated by the British Government show the influence of the British over the State in all the matters. Sir Richard Temple, as Famine Delegate of the Government of India, made a visit to the Hyderabad State in January 1877. In his Report to the Viceroy, Temple said that: 'So far as he was able to judge, the arrangements made to meet distress, and the diagnosis of the coming trouble, were creditable to the prudence and foresight of HES the Nizam's government.' Soon after Richard Temple's visit, the Resident Sir Richard Meade forwarded a full Report to the Government of India on the prospects in the Hyderabad State. In it he said: 'In all his proceedings on this occasion, Sir Salarjung has acted in accordance with the principles enunciated by the British Government, as those by which its own officers should be guided to the districts, under his administration...'; RHF (1876–77, 1877–78), p. 8.

8 *The Deccan Post,* 22.3.1901 and *Imperial Gazetteer,* p. 49. This famine affected the country as a whole. This famine was described by Lord Curzon as the severest and most terrible of all the famines that had affected the country in the nineteenth century. It extended over an area of 1,89,000 sq. miles. Along with the Hyderabad State, the parts affected were, Gujarat, Bombay, Rajputana, Berar, Central Provinces, Southeast districts of Punjab; B. M. Bhatia, *Famines in India,* Asia Publishing House, Bombay, 1963, p. 250.

per cent. In some districts light rainfall averted serious famine. Thus, finally the famine area was found to be the whole of four districts of Aurangabad, Bir, Parbhani, Naldurg and parts of Nander, Gulbarga and Bidar, affecting an area of 23,007 sq. miles, with a population of 3,573,651. The slightly affected area comprised all the remaining districts, except Nizamabad, and extended to an area of 51,541 sq. miles, with a population of 6,512,379. Due to shortage of food grains, the State imported food grains amounted to nearly 3 million maunds to feed the hunger striken population.[9]

By this time, the government got some experience in dealing with famine situation and appointed A. J. Dunlop, a British administrator from Berar, as the Famine Commissioner. He was given a free hand in dealing with relief measures and other activities of the 1899–1900 famine. Arrangements were made for copies of all famine papers circulated by the Government of India to local governments and administrations for guidance and information to be furnished to Nizam's government for the Famine Commission. Bombay Famine Code was adopted in the Hyderabad State and the Famine Commission was working under its provisions.[10] Relief measures were taken in the early part of November 1899 and it continued till October 1900. *Taccavi* advances for the purchase of cattle and seeds, and for sinking wells were given to the extent of Rs. 12,28,616. The total expenditure incurred on the famine relief measures was about Rs. 2 crores.

Local Scarcities/Famines

The distress began with the crop failures in 1899 and it spread in some parts of the State up to 1903, where rainfalls were below the average. As a result some regions met with local scarcities and usual relief measures were adopted. No mortality of any significant nature was reported. During 1906–12 scarcity conditions again recurred and in 1913 there was a severe fodder famine in the Marathwada districts in which many thousands of cattle perished and scarcity conditions and high prices prevailed throughout the western districts. There was practically no fodder left in the State particularly in western parts, as there was no existence of forested areas and extensive grazing lands. The government was compelled to import vast quantities at great expense from distant places in the Tapti valley and from the Bombay Presidency.[11] Finally the last severe famine occurred in the period from

[9] *The Deccan Post,* 22.3.1901.

[10] Ibid., 19.1.1900.

[11] HEH the Nizam's government, *Report of the Famine Relief Operations during 1918–19, 1919–20*, Government Central Press, Hyderabad, (hereafter RFRO), pp. 4–5.

1918–19 to 1921–22. This was caused by the excessive and untimely rains during 1918 and by the heavy rise in prices of things of daily use on account of the World War I.[12] There were 40 *talukas* that were partially affected and the total area declared as famine-stricken was 12,643 sq. miles with a population of 22,37,529. The famine coupled with heavy attack of epidemic diseases lead to a huge loss of human and cattle life due to shortage in food and fodder. In fact, the mean density of population per sq. mile in the State came down from 13,374,676 in 1911 to 12,471,770 in 1921 with the density from 162 to 151.[13] Agricultural depression was characterized by rising prices of food grains.[14] Relief measures were started in 1919. A Central Famine Committee consisting of ministers and secretaries of important departments was appointed by the government to control famine operations. Mohammed Ali, Director General of Police and Jails, was given the charge to act as the Famine Commissioner. On his recommendations the famine relief works in the State were carried out on the lines laid down in the Hyderabad famine and the Bombay famine codes rules and the provisions contained in these two codes therefore governed all the operations.[15] PWD relief works and village relief works were started as per these codes. Poor houses were established for those who could not work on account of old age, ill health or infirmity. Arrangements were made to afford gratuitous relief to those who would not join relief works or for other reasons. In addition to this *taccavi* loans were given to the farmers for sinking and repairing wells, for the purchase of seeds and cattle, and for weeding fields. Kitchens were started in famine relief camps to provide food to those dependants of labourers who were unfit to do any work. When local supply failed, food grains and fodder

---

12 The average rainfall of all the talukas of the marginally noted eight affected districts was 40.22 inches while the normal was 34 inches (the 8 districts were Aurangabad, Bid, Medak, Mahaboobnagar, Nalgonda, Warangal, Karimnagar and Adilabad). In 1919, the rainfall was much below normal being only 22.54 inches; RFRO (1918–19, 1919–20), pp. 1–3.

13 There was a decrease of 9,02,906 persons in the decade of 1910–20, shows the intensity of this famine. *Census Report of 1911*, Hyderabad State, pp. 15–16 and *Census Report of 1921*, pp. 33–34.

14 The price of jawari during 1919–22 period ranged between 5.5 and 3.5 seers in the State and 4 to 2 seers per rupee in the famine-stricken areas. The dearness of foodstuffs in this famine years intensified suffering to an extent that had never been experienced in previous famines. *Administrative Report for 1329 Fasli (1919–20),* Government Central Press, Hyderabad, 1921; *Annual Average Retail Prices of Principal Commodities in HEH the Nizam's Dominions, 1919–50 Fasli (1921–1941),* pp. 6–93; and RFRO (1918–19, 1919–20), pp. 9–10.

15 RFRO (1918–19, 1919–20), p. 7.

were imported from British India and cheap grain shops and depots for fodder were set up in places where needed. Medical aid was provided and efforts were made to save the farmers in the affected areas from the effects of epidemic diseases.[16] On the whole the relief measures taken by the government could only lessen but hardly mitigate the suffering of the rural population.[17] In both 1900 and the 1918–22 famines the districts worst affected were those along the western border. In 1927 again scarcity conditions existed in Lingsugur and Sindhnur *taluks* of the Raichur district and in parts of Nalgonda district. In 1934–35, parts of the Raichur, Gulbarga and Karimnagar districts were affected and an expenditure of Rs. 17.03 lakhs debitable to the Famine Relief Fund was incurred.[18] On the whole, people during this famine were in a better position than the earlier famines to resist distress. A large number of persons could migrate to safer places and find livelihood there.

The State experienced a number of scarcities but no major famine over a period of around 20 years, from 1920 to 1940. The scarcities did not involve considerable loss of life. It was during 1940–42, when the State was confronted with a famine, which officially was not a famine but one of scarcity conditions, the famous Bengal famine was on its way to rout the people. The scarcity was a result of a series of crop failures that the State was unable to meet since 1938.[19] Outbreak of the World War II in 1939 aggravated the situation and affected the country in many ways. The total amount of rainfall that was received in respect of the State was below the minimum required level resulting in a huge loss in respect of total area cultivated and the outturn received. Normal imports of food grains were ceased and wartime controls dislocated private trading and movement of food grains on account of provincial and even district barriers against the movement of grains. Natural shortages and dislocations in the normal channels of distribution and the resulting tendency on part of traders and consumers to hoard, shot up the prices to abnormally high level. Though the State suffered to a lesser extent, however, prevalence of famine in other parts of British India had a serious effect on its day-to-day life. Government took relief measures and was slow

---

16 Ibid., pp. 2–4.

17 Though the Famine Report of 1918–20 boasted of its measures, the ground reality was assessed by the General Administration of the State, see *Administrative Report for 1330 Fasli (1920–1921),* Hyderabad, 1922, p. 1.

18 HSFM, pp. ii–iii.

19 Warangal Suba file No. 58 of 1350 Fasli (1941) and Record No. 2506 of 1351 Fasli (1942); *Trade Statistics,* HEH the Nizam's government, 1351 Fasli (1941–42), pp. 12–13.

in execution of relief programmes, which increased sufferings of the people in affected areas. But remission and relief measures were hardly able to compensate for the loss. Again in 1945–47 the State received insufficient rainfall (only 28.85 inches) causing failure of crops. Though there were sufficient rains in 1948, the State was experiencing political and economic convulsions that had an unsettling and damaging effect on crop conditions. It is, however, significant to note that though failure of rainfall was responsible for producing scarcity conditions during the last 3 decades from 1920s yet, the conditions never deteriorated to such an extent that they approximated to the horrifying years of famines before 1919–22.

Famines in Retrospect

It is clear from the above discussion that the predominant cause of famines in the Hyderabad State had been rainfall either excess or shortage. Rainfall disturbed agricultural production affecting initially the *kharif* crops which provided about half the staple food-grains of the people.[20] If it was followed by an absence of autumn rains, the *rabi* crops as well as *jawar* crops also suffered. This can be amply illustrated by the fact that in all the famines that affected the Hyderabad State in the nineteenth century, *jawar* was extremely expensive and beyond the reaches of the poorer classes. It had been the failure of *kharif* crop that had led to increase in their prices and ultimately beyond the purchasing capacity of the poorer classes leading to the occurrence of famine.[21] But, failure of croping, especially *kharif* crop, created famines in the *famine zone* more easily than in other areas of the State. The explanation for this lies in the geographical position of these districts of Marathwada which affected the rainfall and also the possibility of creating irrigational facilities. The Hyderabad State was not endowed with natural lakes and irrigation was generally carried out by means of artificial lakes. The Pakhal Lake was one such example. The total number of tanks in the territory was 18,199 of which 3,000 tanks were in Khammam district, 1500 in Karimnagar, 2500 in Nalgonda and 2000 in Nizamabad.[22] The often famine stricken Shorapur district had a few tanks across the tributaries of the Krishna river. Geological conditions of the Marathwada region did not permit irrigation to be carried out on a large extent. Mahadi Ali's report on famines illustrates the clear distinction between the Marathwada and Telangana districts of the

---

20 *Imperial Gazetteer,* p. 49.

21 RHF (1876–77, 1877–78), pp. 20–25.

22 *Report on the Administration of HEH the Nizam's Dominions for the year 1331 Fasli (1921–22)*, Companion Volume, pp. 2–3 and 95–96.

State. Artificial irrigation was more common to Marathwada region than Telangana region as the soil in the Marathwada region was porous and bunds got cracked and fissured in hot season, and therefore often got easily breached during the rainy season. Further in the Telangana districts the soil was less fertile and rivers generally would go dry, except in rainy season and hence the necessity for the collection of water supply. In contrast to this, the Marathwada region was endowed with fertile soil which retained moisture.[23] Irrigation was largely unknown in the Marathwada region when Mahadi Ali furnished his report on the famines in 1876–77 and 1877–78. In later period also the same situation continued. Mahadi Ali further noted the necessity of introducing a system of wet cultivation in the Marathwada region by a system of well irrigation or by construction of masonry tanks. This was accentuated by the fact that monsoon rains were highly unpredictable in this region and artificial irrigation was necessary. The nature of the soil in Marathwada region being porous needed less labour for crop production as compared to the regions of Telangana which would come under wet cultivation.[24]

The soil in the Hyderabad State therefore determined the extent to which cultivation can be carried out. With dependence on irrigation either by wells or tanks, rainfall would become an important factor affecting crop production in the State. The average rainfall during the 50 years period (1895–1945) in the Hyderabad State was 33.73 inches; out of which the average of the Marathwada and the Telangana was 28.53 and 32.92 inches respectively.[25] The Hyderabad State, especially those regions of Marathwada was predominantly dependant on summer rains. Owing to their geographical position it depended almost exclusively on the summer rainfall and winter rains were almost negligible in this region. But the eastern and southern portions of the State were also nourished by the autumn rains, when the same currents were deflected on the eastern coast. It might also be expected that the distribution of rain would vary according to the distance from the coast, but in descending over the Western Ghats, the portions immediately to the east of these hills would receive less, while the increase was only gradual. There was, however, a limit to this increase, and as the country falls in the valleys

---

23 RHF., p. 33.

24 *Report on the Administration of HEH the Nizam's Dominions for the year 1331 Fasli (1921–22),* Companion Volume, p. 10.

25 *Decennial Report on the Administration of the HEH the Nizam's Dominions, 1912–22,* Government Central Press, Hyderabad, p. 16; S. M. Bharucha, *Agricultural Indebtedness in HEH the Nizam's Dominions,* Hyderabad, 1937, p. 24; *Statistical Year Book,* 1935 and 1942–45, pp. 294 and 881; 1941, p. 529.

of the principal rivers. The upper currents veer towards the eastern coast and constitute the winter monsoon. Hence the rains of the summer monsoon were quite general, though not equally distributed throughout the country.

The Marathwada region was barren of forest vegetation and marked by absence of large areas of irrigated lands. Summer rains were more in this region than those of other regions of the State. Telangana region which was endowed with large forest vegetation, though away from the coast, however, managed to get enough rainfall due to the fact that '...a net work of tanks which support extensive areas of wet cultivation is found throughout the country, and, from the effects of the evaporation that is constantly going on, there is a considerable amount of moisture in the atmosphere which reduces the temperature and favours the precipitation of rain from the currents that pass over. Hence, although Telangana is further removed from the western coast than Marathwada, the difference in the rainfall during the summer monsoon is not great; while during the winter rains the excess is in favour of Telangana from its proximity to the eastern coast, and, on the whole, the annual fall of rain is greater in Telangana than in Marathwada.'[26] However, it is to be noted that the atmosphere in the hot weather in the Telangana was not so dry as in the Marathwada. The Telangana was, as said in the above quotation, studded with a large number of water tanks and reservoirs for irrigation purposes. The large forest areas in the region would make the air somewhat humid in the hot season.[27] Crop production was also dependant on the type of soil available. The soil of the Marathwada region facilitated cultivation such as cotton, wheat, jawari, bajra, etc. Only after 1900, efforts were made to provide well irrigation in Marathwada and the construction of big dams in the Telangana reduced the severity of famines.[28]

It has been observed that the net area sown under cash crops was continuously increasing which became one of the causes for the prevalence of famines from 1930s onwards. For example, A. I. Qureshi has shown the increase in the area under commercial crops from 1930s onwards. He has given the statistics of castor which had 54.3 per cent of the total Indian acreage, 15.6 per cent of the total acreage under groundnut and 90 per cent of which was sown in the Telangana region.[29] The statistics of Agricultural Department during the year of 1352 Fasli (1942–43) also shows the fact that cotton and

---

26 RHF, p. 36.

27 Department of Agriculture, HEH the Nizam's government, File No. 35, Serial No. 97 of 1357 Fasli (1948).

28 *The Deccan Post,* 1.2.1901.

29 A. I. Qureshi, *The Economic Development of Hyderabad,* vol. I, Orient Longman, Bombay, 1947, pp. 53–54.

groundnut area increased over 11 and 15 lakhs of acres respectively. But, it also shows that the area under the food crops considerably declined. For example, the decline under the *jawar* crop was about 1 lakh acres and 3 lakh acres under wheat.[30] In the State, cash crops were cultivated in 75 per cent area and only in 25 per cent area food crops were sown during 1940s. Many complaints were lodged against the cultivation of cash crops during the time of famines for the selfish needs. People strongly objected to this type of cropping pattern and began to send requests to reduce the area under cash crops. The state government issued a warning that all the farmers should reduce their cash crops to one third of food crops. It gave some measures in 1944 that mixed cropping pattern should be practised, consisting of 2 parts of food crops and 1 part of cash crops. It cited the example of the Bombay Presi-dency where food crops occupied 2 parts and cash crops grew in 1 part out of the total agricultural land.[31] It passed an Act on the cropping pattern in 1944 to restrict people to grow food crops on the lines of the Bombay Presidency. However, due to the political disturbances the Act might not be implemented properly.

Dependence of People on Forests during Famines

Local people, both the peasants and the tribals, depended on forest products during the periods of scarcity and famine. Those who could not afford buying food grains at exorbitant rates during famine periods, naturally sought alternatives or substitutes and the ideal grounds for them were forests. During all the famines the poorer classes were the most affected people and one indication of this affliction was an increase in the dependency on forest products like mohwa flowers, achar, tendu leaves and tamarind seeds. The forest resources were of critical importance to them not only for the nutritional values but also for the economic, cultural and aesthetic values encompassed within hunting, grazing, fishing and food gathering.

In the places near to villages, fruit-bearing trees such as mahua, chirongi sitaphal, neredu, jackfruit, ber or regu, mango, edible berries, tamarind, etc.[32] were grown. The mahua tree was found all over the State and was very important to the local economy, as it was also the most important food resource of tribal people. They used to consume all products of mohua tree, i.e., flowers, fruits and seeds. There were many other varieties of forest

---

30 *Hyderabad Samacharam,* vol. 4, no. 11, 1946; *Meezan,* 14.6.1944, 19.7.1944 and 9.9.1944.

31 *The Golkonda Patrika,* 18.5.1944.

32 CVF. Haimendorf, *Tribes of India,* p. 80.

plants and trees that yielded fruits which were taken as valuable food supplements during famine periods. In addition to these fruits, leaves and barks of some plants like tendu and achar were used as vegetables. The tribals were heavily dependant on forest products. For example, Gonds and Kolams of Telangana region were dependant on the forest fruits, leaves, barks, tubers, honey, etc.[33] Local people and forest tribes were fully aware of forest environment and its surroundings and knew much about the season and the availability of fruits, nuts, leaves and tubers grown there. They had the excellent knowledge of the nature of wild fruits, whether they were edible or not. The skill of the local people and tribals in collecting these wild forest foods and their knowledge of the plant and animal kingdom protected them in times of famines. These wild forest foods were eaten by tribal people as a staple food, but local people used them only in times of scarcities. The tribal population were less affected by famines due to the availability of forest products, when compared to the people dependant on agriculture. Availability of edible items in the surrounding areas minimized the risk of drought.[34] This might be the reason why the Marathwada region was more affected by the famines than the Telangana region in which the existence of large extent of forest tracts minimized the incidence of famines and consequent food scarcity.[35]

Till the establishment of the Forest Department, the peasant and tribal communities pursued relatively undisturbed lives in and around the forest areas. However, with the commencement of the forest management by the State their freedom was restricted. Further, larger economy under the British colonial influence increased pressures particularly on the tribal communities over their living and ended their relative isolation. With the establishment of the Forest Department, the Hyderabad State gradually imposed its authority over forest tracts as well as on forest-dwellers and occupied all the forest rights, which came down to the forest people since time immemorial. The government embarked on a wholesale programme of reservation of forests and exclusion of forest-dwellers.[36] Further, the state administration put forth

---

33 K. Mohan Rao, *The Kolams: A Primitive Tribe in Transition,* Booklinks Corporation, Hyderabad, 1990, pp. 19–20.

34 Ibid. p. 19.

35 The severity of famines was more in the Marathwada region prior to the beginning of twentieth century due to the absence of large scale forests. In the twentieth century, the incidence of famine deaths was reduced and the conditions were improved because of the development in transportation. Foodgrains now could be sent to the affected areas in less time.

36 K. Mohan Rao, *The Kolams,* p. 20.

an argument that the traditional forest use seriously exacerbated the destruction of forests. It was said that peasants were continuing age-old practices which had scarcely harmed the forest when it was in abundance. Felling a tree to make a plough or house construction was regarded as a wasteful habit of local people. This official antipathy towards people's habits can be clearly observed in the comments of W. F. Biscoe, Conservator of Forests, that 'In some forests ...many thousands of goats had lately been browsing; 90 percent of the trees had been pollarded by Dhangars. The forests looked as if a cyclone had passed over them.'[37]

In addition to State intervention, the influential classes of the State such as *jagirdars, deshmukhs, deshpandes, patels* and *patwaries* illegally occupied all wastelands including forest and pasture lands and began to treat them as their own. They began to charge people to use the natural resources. All the customary rights of the people were denied and treated as illegal. Thus, forest-dwellers and peasants were deprived of almost all their forest resources and common rights by the activities of the State and the rapacious feudal lords.[38]

During the major famines, effects of economic and ecological changes were felt seriously in western part of the State, where the lands were parched and forests were dried and deteriorated. Reporting on the growing scarcity, R. M. Crofton, Director General and Revenue Secretary, observed that the degree of intensity of the famine would vary with regard to the availability of forest produce. He, therefore, identified the areas most likely to be affected as the districts in the western part of the State i.e., Marathwada and Karnataka districts followed by the Telangana region.[39] During the famine of 1876–78, all the districts in the Marathwada region were reported to be famine-stricken and it was recorded that the sufferings of the people were great. Many people were dependant on available forest products such as leaves, tubers, seeds, wild fruits, etc. in addition to the limited relief measures offered by the State.[40] 'People knew to some extent how to combat starvation during famine periods' by consumption of natural resources. However, with the onset of the state management of forests the traditional economy could no longer with-stand the distress of famines. As many of the forest areas were under

[37] ARFD, 1308 Fasli, (1888–89), p. 9.

[38] CVF. Haimendorf, *Tribes of India,* pp. 51–57.

[39] HSFM, pp. 1–2.

[40] The *talukdar* (district collector) of Hyderabad gave a note that due to shortage of food grains and the high prices made people to live on the leaves of trees during the famine of 1876–78; RHF (1876–77, 1877–78), p. 23.

the control of the State and many landlords held private forests extensively, common people did not have free access to the forests as was earlier to them. However, government opened some of the forest areas for public needs during famine periods for collection of forest food and fodder for livestock. Though the government wished to open all types of forest areas to alleviate the distress, it was delayed by the Forest Department. Its main objection to the proposal was that the public access was liable to cause forest fires, as it was a custom of the people to clear the land under the fruit-bearing trees by setting fires. The Forest Department used to catch the people who set fires and performed 'prohibited' activities and this could be evidenced by the increase in the number of forest offences during famine periods.[41]

Pressure on Forests

The strata of the population most severely affected by scarcity were those with no land holdings and agricultural labourers, i.e., low castes and tribals. Smaller land-holders, landless classes such as agricultural and general labourers and petty artisans were the sections, who, having no stores of grain, were dependant on what they could earn by daily wages. They were the sections of society mostly affected by famine.[42] Retention of traditional rights to forests, pasture and other natural resources was important to the subsistence economy of the poor peasants. It was reported that whenever food grains were over, the women were foraging far and wide for edible plants. In most of the cases men migrated in search of work, while the women and children were left to forage for food. Forests primarily served as a principal source of living particularly at the time of scarcity and famine. The minor forest produce which included a variety of nuts, fruits and tubers, leaves, seeds became the source of survival. But due to Forest Department's strict restrictions and the reservation of forests many people had been deprived of their pasture and forest rights. It all depended upon the mercy of petty officials of the Forest Department.[43]

During famine periods, as food becoming costlier and relief measures being inadequate, the government sometimes allowed the opening up of some forests as the part of famine relief. This incursion of the people from the plains into the forests was not a smooth task, as it had devastating

---

41 ARFD, 1900–1901, p. 7.

42 HSFM, p. i; *The Deccan Post,* 1.2.1901.

43 CVF. Haimendorf, 'The Problem of Shifting Cultivation in Hyderabad', in *Hyderabad Forest Magazine,* Vol. II, No. 2, 1942, pp. 1–8. And *Tribes of India,* pp. 79–96.

ecological implications.[44] In other words, dependence and pressure on forests clearly enunciated the feature of an agrarian economy for survival and to meet even the daily requirements. Forests, therefore, in times of famines became a contested site amongst the Forest Department, locals, tribals, revenue officials and a large chunk of population living on the fringes of the forests. The disturbance in the agrarian economy to sustain itself during the famines led to increased dependence on forests.[45]

Apart from providing people with direct dependence on the forest produce during the time of distress, certain minor forest produce was also supplied to the government for relief works. Often government relief works were carried out by expanding the activities of the Forest Department, such as climber cutting, cutting back, thinning, etc., and opening up new works such as road improvements and new constructions to provide employment in the affected areas.[46] For these employment generation works, funds were allotted by the Inspector General of Forests.

Fodder Scarcity

Forests acted as a direct source of dependence in times of famines not only to human populations but also to cattle by providing fodder. Unfortunately sometimes fodder scarcity was closely followed by famines. For example, in 1937, the border districts of Aurangabad, Bhir, Osmanabad, Gulbarga and Raichur suffered from a severe fodder scarcity and large numbers of cattle died for want of fodder.[47] Cattle need grass as well as water to digest the dry grass. Accounts of cattle deaths as a result of scarcity of water in the grazing

---

44 In the famine of 1899–1900, many fires occurred in the January 1900 and there was scarcely any grass left for the vast herds of cattle which had come to the Northern Telangana districts from the Marathwada region in search of fodder. But unfortunately many thousands of cattle died which could have been saved, had the fire protection on a large scale been in force that year. The Conservator of Forests, W. F. Biscoe observed that people set fire to forests for their selfish ends which caused enormous damage to the environment and would show adverse effects during the time of successive failures of rains. He suggested to the Famine Commissioner that grass should be cut and stacked in the early season; this practice though had been in practice from the last decade of the nineteenth century, was not implemented due to lack of funds and official disinterest. In addition to it, people also suffered for non-availability of forest foods as there was nothing left or was not grown any thing due lack of rains; *The Deccan Post,* 14.12.1900.

45 *The Golkonda Patrika,* 1.9.1941.

46 *The Evening Mail,* 11.5.1899 and 14.11.1899; *The Golkonda Patrika,* 1.9.1941.

47 Ibid., 68. *The Golkonda Patrika,* 13.9.1937, 30.12.1937 and 25.4.1938.

reserves inside the forests were often reported.[48] To avert this, the government made efforts to construct temporary dams across the *nallas* (channels). At certain times owing to the prevalence of scarcity conditions within the grazing reserve and outside as well, construction of *nallas* was undertaken as as one of the provisions of employment for labourers and this was included within the relief programme. To tide over acute fodder shortage, import of fodder from other parts of the State or outside was also accepted. Traders were notified of the relief tracts where there was fodder shortage. A famine protective policy was formulated to prevent a fodder shortage particularly in the famine zone where rainfall was less. For example, as a result of the 1937 fodder scarcity, the construction of the Rooty and Sakat Tank projects were sanctioned. Each of these projects was inclined to provide permanent solution to overcome the fodder scarcity in Ashti and Parenda *taluqs* and also provide enough fodder for about 10,000 cattle for six months of the year. Another remedy to tide over fodder shortage and prevent the huge loss of cattle life was to send them to areas where fodder had already been stacked at a cheap price. In 1937, some 20,000 cattle were sent to famine *rumnas* in Aurangabad and Bhir districts and other cattle to Nanded districts where it was proposed to buy up and stack *kadbi* (*jawar* or *bajra* stalk) near grazing grounds.[49] Control over movements of cattle also formed one of the important measures to alleviate fodder shortages. For example, in the 1937 fodder scarcity, the State came under heavy pressure by a large influx of cattle from the adjoining territories of British India. To cope with this kind of problem in future, special precautions were undertaken to guard against such migration of cattle into the State by charging grazing tax, etc.

Storage of fodder and their transportation and distribution were also introduced to avoid fodder scarcity. As a result of fodder scarcity in 1940s, grass bundles were imported and kept in fodder depots like in Nalgonda district. It was utilized in bailing the grass during current and future famines. Efforts were also undertaken to store 50 lakhs pounds of grass at the famine *rumnas* in order that a nucleus may always be available to depend upon during the scarcity conditions.[50] As the famine zones had only 1.24 per cent of the total area of the districts under forests, the remedy therefore suggested was to supply it with fodder from other areas of the State. This necessitated

48 To cope up this problem Well-Sinking Department was established in 1933 as a relief measure; *The Golkonda Patrika,* 18.3.1937. One of the objectives of the construction of Nizam Sagar Project was also to reduce the distress caused by famines; *The Golkonda Patrika,* 20.7.1932.

49 HSFM, p. 136.

50 *The Golkonda Patrika,* 1.9.1941. *Meezan,* 14.6.1944 and 7.10.1944

a good network of rail and road links that would facilitate the procurement of fodder from other areas into the forest zone. The railway extensions from Parli via Bhir to Aurangabad and from Nanded to Adilabad were contemplated, as it would then be possible to move fodder by rail right into the heart of the affected area.[51] In 1937, as there were no direct rail links, arrangements had to be made for the transport of one crore pounds of grass from the Nizamabad and Nanded districts by rail to Jalna and from Jalna by Nizam State Railway buses to Sonthada in the Bhid district.[52] As the cost of road transport was heavy, other measures like better storage facilities, fodder remissions were effectively enforced to tide over scarcity conditions. However, the steps undertaken to alleviate distress were proved to be grossly inadequate.

The form of relief afforded from the forests during famines varied with the locality and the rules in force. Also, the nature of permission provided to remove forest produce either free or at concession rates to supply fodder and the expansion of normal forest works.[53] The throwing open of the forest resources to the poorer classes was often recommended by the government and famine relief administration, but it was not heeded to the desired extent by the Forest Department. It imposed several restrictions on the use of forest resources. Some of the rules were:

1. removal of dry fuel and grass only in head-loads allowed
2. free removal of limited edible fruits
3. grazing dues not to be collected from agricultural cattle and grazing fee collected from professional graziers and
4. throwing open only a limited 'closed' coupes to grazing.[54]

The measure of 'throwing open of the forests' was often considered having underlying commercial considerations. Economic considerations and usual balancing of vested interests remained the pivot around which peoples' needs were allowed. But strangely ecological considerations were not taken into account. This precipitated deforestation on a large scale and the loss of valuable pasture. On the otherhand, government revenue and established agencies tightened of their grip over the forests to get economic benefits when people were struggling for their subsistence.[55] Restrictions on the use

---

51 Ibid.

52 Ibid.

53 HEH the Nizam's government, Forest Department, Instalment No. 16/75U, File No. 1/30, Rec. No. 547/1351 Fasli, 1941, (IRH).

54 HSFM, pp. 64–66.

55 The government had taken full consideration in terms of safeguarding state revenue and private contractor's interests. Because if any 'closed coupe' declared open for famine

of edible fruits that could be removed by the poorer classes were also specified. Only certain species like mahua, achar and tendu were allowed to be used. Relief or remissions to the poor offered by the Forest Department during the ties of scarcity was often subservient to the contractors' interests.[56] Removal of dry fuel in head-loads at concessional rates or for free was not permitted inside the forests. But it was free only on the fringes of towns and villages where poorer classes depended on sale of head-loads of fuel for their living. The mobility of the poorer classes inside the forests was also restricted. Whenever the concession to the poorer classes was offered it was tagged with the stipulation that the transfer of head-loads by carts within a reasonable distance of the forest was not permissible. Only during the severe famines such as 1876–78, 1899–1900 and 1918–20 famines, all types of forests were thrown open to grazing, fuel and food collection. For example, during the famine of 1876–78, people were allowed to collect forest food and the cattle were permitted inside the state forests for foraging. During the famine of 1918–20, all the reserved forests in the districts of Warangal, Nalgonda, Mahaboobnagar and Adilabad were opened to the farmers of the famine-stricken areas to graze their cattle free of charge. Cattle were also allowed to graze in government *rumnahs* in the Marathwada area, under the supervision of a special officer of the Forest Department.

Pressure on forests was also added as the Forest Department made the supply of forest products to famine relief centres. Poles, grass and other hutting materials and firewood were supplied free of charge to state relief works. However, it kept some restrictions on the relief works regarding the distance of a relief camp from forest areas and the selection of coupes to prevent injury to the forests, etc. Though the State and the Forest Department allowed people to use forest resources free of cost, many petty forest officials resorted to corrupt means and cheated the illiterate and innocent poor people

---

relief operations, the government had to pay compensation to the coupe contractor for the loss incurred due to people's access to it. The government took utmost care to 'minimize' the loss to the state exchequer by making strict vigilance over people's action in those areas.

56 The contractors' interests were also protected in procurement of grass. The poorer classes, who were offered concessions, though allowed free removal had to compensate the contractor before cutting the grass. Their expenditure consisted of kists paid, the wages of the watchers and money spent on cutting and stacking. Sometimes the government also paid to contractors on behalf of people for allowing free grazing. HSFM, p. 64. As a part of famine relief measures, in Maharashtra rumnahs (pastures) were opened up for grazing by giving compensation to *Guttedar* (contractor); *The Golkonda Patrika,* 19.10.1936.

taking *mamuls* (bribes).[57] Many reports had come to the notice of the contemporary newspapers[58] that the Forest Department staff did not stop collecting grazing and other taxes during the periods of scarcities.

Thus, the present chapter has discussed in detail about the occurrence of famines and people's dependence on forests during the famine periods. It has been observed that, people of this region had traditionally evolved their own survival strategies based on their prevalent natural resource base. Peasants and tribal communities used to depend on forest resources during famine and scarcity conditions. Forests in particular used to provide traditional support in absorbing the shock of natural calamities like famines. However, with the interference of the state government these customary rights and traditional strategies were curbed and made dependant on the mercy of the State.

It has been observed that though the causes of famines were largely due to failure of rains, the severe impact was due to policies of the state government like taking control over forest resources, grain exports to British India and Europe, low wages and high prices. The state government and its departments sometimes refused to acknowledge the existence of famine or scarcity in this region. This led the bad situation to worse, as observed by the contemporary newspapers. The inadequate relief measures and indifference of state administration towards public spending resulted in popular discontent and large-scale deaths. Incidents like Babhijheri and famous Telangana peasants' armed struggle were the results of such discontent amongst the tribal and peasant communities.[59]

---

57 There were so many instances of corruption at lower levels; see *The Golkonda Patrika,* 10.8.1939, 15.7.1940, 31.7.1941, 15.9.1941, etc.

58 *Deccan Post,* 2.2.1900, *The Golkonda Patrika,* 10.8.1939 and *Meezan,* 13.1.1946.

59 The policy of clearing large tracts of forest of all human habitation, including old and established villages inhabited for many generations, led to the only case of armed resistance in 1940 by the Gond tribal community in Adilabed district under the leadership of Kumaram Bhim at a place called Babijheri. For details see, CVF. Haimendorf, *Tribal Hyderabad,* Hyderabad, 1945, p. 122. D.Venkateshwara Rao, *The History of the Telangana Armed Struggle,* Hyderabad, 1988, p. 133.

CHAPTER 7

Conclusion

The formation of the Forest Department in 1867 and the policies it enunciated had a far-reaching impact on the social, economic and ecological aspects of the Hyderabad State. Forest vegetation of the Hyderabad State was a part and parcel of the great deciduous belt occupying a large portion of the peninsular India. Till the middle of the nineteenth century, the State did not develop a 'commercial attitude' towards natural resources like forests. As elsewhere in pre-colonial India, rulers considered the right of the peasant is and tribal communities to utilize forest products. As the population was low, demand for forest recourses was also moderate and export of forest products was non-existent. There was availability of fuel, fodder and non-wood products in abundance for domestic consumption by local communities. It is generally argued that there was no significant evidence of conflicts over control of forest resources.

It is in this context that the British government introduced changes at the all India level during the nineteenth century in property relations and state restrictions on the use of natural resources by tribal and peasant communities. With the construction of railways during the 1850s and 1860s, there was an increased need to meet the demand for railway sleepers and fuel and it is in this context that the Hyderabad State felt the importance of forests. The idea of inexhaustible forests gave way to the search for suitable strategies for control, conservation and commercial exploitation of forests. Against to this background, Salarjung I established the Forest Department on 16 July 1867 during the rule of Afzaluddaula the Nizam V, by following the British model. Initially, the state administration took some measures to protect some valuable tree species for fuel and timber and for their regeneration and development. However, there was hardly any investigation into people's customary rights that were in practice prior to this period. The assumptions of the British administrators were that the local populations had never enjoyed any rights and that they were by nature, destructive, and therefore, were to be stopped. This assumption was uncritically accepted by the Nizam's Hyderabad State. It is significant to note that the three important and key branches of administration, i.e., revenue, finance and police were kept under the control of British civil servants, which naturally pointed to a strong British influence over the state administration in Hyderabad. This was not new to the princely states like Tehri Garhwal, Tripura, Mysore and Travancore which had also established their Forest Departments under the broad policy framework of British forest management.

The first task of the Forest Department in Hyderabad State was to check reckless deforestation and identification of the sources of strong and durable, commercially important timbers such as teak. Protection of forests began as a response to the depletion of large size teak trees in the forested districts of Telangana. Initially it was largely a question of enforcing a minimum size to be cut. Timber trade had been a prominent item of income from the middle of the nineteenth century. There was then little concern about its depletion. In the beginning, simple rules were framed for regulating the lopping of trees for railway sleepers and fuel, while trade by people in any form of forest product was prohibited. In practice the protection to timber trees was more crucial than the half measures like establishing government control on timber-producing land. The effectiveness of protection for valuable timber was limited because only a small establishment in its initial stages could not do much in controlling extensive forest tracts.

In other words, the infant Forest Department and its few personnel could hardly be considered effective in conserving the forests in Hyderabad State. During the first two decades of the formation of the Forest Department some valuable species were declared 'reserved' trees and the primary activity of the Department was to maintain and protect these species. The management of rest of the species, forest produce and administration remained in the hands of the district revenue officials. During this period the state also launched new schemes such as *Ijara* for the expansion of cultivation. New rules to repopulate deserted villages were brought into force. Subsequently lakhs of acres of forest area was reclaimed and passed into the hands of private individuals or peasants.

Initially preservation of existing forests was the *raison d'être* of the Forest Department, and survey and demarcation of forest lands were carried out to conserve them. Later, the whole policy of conservation was pursued with the object of facilitating the production of timber, both for revenue and for British imperial purposes. Gradually, the Forest Department established itself as a fully commercial department, without being sensitive to the needs of common people.

Though there were forest policy initiatives in the nineteenth century, the first Forest Act was passed only in 1900, which was later modified, on the complaints from the foresters, and enunciated as the Forest Acts of 1917 and 1946. These acts divided forests into three main categories: reserved, protected and open forests. As in British India, Forest Acts of the Hyderabad State gave power to the Forest Department to decide how forests were to be managed.

The state had always given preference to conservation in its policy statements; however, in practice it could never achieve the goal set by the department itself. The forest policy of the Hyderabad State, which was enunciated in 1890, made it clear that the over-riding consideration would be 'public interest', which in practice was interpreted as the interest of the larger state economy. Where the state interest was in conflict with the local interest, the latter had to be subordinated to the former. Naturally, there was discontent among people. The exploitation of forests for commercial purposes was claimed to be of 'public interest'. But cultivation of arable lands, that was shown as forest lands but without any actual tree cover, was treated as encroachment. Ballantine, an experienced British forest officer, on his arrival in 1887 as the Conservator of Forests to the Hyderabad State remarked that the Forest Department had been useful as a 'collecting agency and had failed in matters relating to conservation and reproduction'. Even the Revenue Department in its report of 1914–15, criticizes the overt commercial orientation policy of the Forest Department.[1]

With the state control over forests, new techniques, strategies and management were evolved. The takeover of large tracts of forestland made several new changes within a short span of time, in the pattern of settlement, agricultural production, agrarian extension, distribution of wild animals, etc. Another important change was that the government tried to claim property

---

1 *The Revenue Administration Report* criticized that 'A Forest Department has been in existence for the past 48 years, but has chiefly occupied itself with the production of revenue and some degree of protection.' HEH the Nizam's government, *Revenue Administration Report: 1914–15*, p. 135.

rights over common grazing lands and forests. The grazing rules came into being in 1894 and were revised in 1912 to make them more burdensome. These rules continued to be unpopular with the common people. In contrast, in the nineteenth century people enjoyed relatively free use of wastelands for grazing purposes. Though, the state had, in some areas, made over some forested parts under settlement to the villagers to utilize as village forests, the loss of community ownership effectively broke the link between local communities and the forest.

Even after taking over all major forests and categorizing them into reserved, protected and open forests, the state had not achieved the target figure of one-third of total geographical area under tree cover. As statistics reveal, by 1914–15 the forest area under the control of Forest Department was 10,762.41 sq.m., which was 13.43 per cent of the total area of the state covering 82,698 sq.m. This, however, was reduced to 11.65 per cent in 1944–45. In contrast, in British India around 20 per cent of the area was under forests before independence. Many schemes and policies were proposed to increase the area under state forests to 20 per cent to the least, but without any positive results.

The government always deprived the tribal communities of their rights on forests, blaming them responsible for the deterioration of forest resources. The state did not make any effort to develop local institutions to manage the forest resources with the participation of communities. Forest people too lost their interest in forest that provided them with their livelihoods. On the otherhand, the state continued to use them as open access resources, which resulted in total degradation or destruction. Many recent studies show that foresters traditionally developed a commercial approach to forest management and ignored the rights of locals, as a result, alienating them.[2]

Although it was officially declared that the policy behind the system of management of forests was 'to conserve, maintain and improve the existing forest areas and working them with the greatest possible care and caution under technically sustained yield management by first and foremost supply the local agricultural and domestic village requirements at cheap concession rates or free supplying the forest requirements within the State.'[3] Though the policy sounds ideal, it largely remained on paper. Moreover the policy did

---

[2] N. S. Jodha, 'Common property resources and rural poor in dry regions of India', *Economic and Political Weekly*, 5 July 1986.

[3] HEH the Nizam's government, Revenue Department (Forest Branch), Instalment No. 219, 220, File No. 131/86, Record No. 420/1350 Fasli, 1941, (Interim Repository, AP Secretariat).

not spell out the organizational reform needed to safeguard local community interests, at least where they did conflict with public interest. It did not suggest any decentralization of control in the form of giving over minor forests to the village or to the community administration at the local level.[4] The state administration did not take up the production of minor forest products such as medicinal plants, spices, honey, etc. This indicated that the state, which reserved for itself the greater part of the profits from the exploitation of timber, did not try to make any serious efforts to develop a local economy around the minor forest products.

However, the pressure from the local communities was such that the Forest Department did give some concessions to peasants in some forests. The Forest Act of 1900 provided for the constitution of village forests which were allotted in the areas of open forests, the rights of people were protected and they exercised privileges in the way of grazing cattle and collecting timber for fuel, and other agricultural or domestic purposes. In practice, the people's rights of access to forests was restricted to limited areas such as 'minor' or 'unreserved' forests, and these areas inevitably and soon became overexploited. The minor forests faced an ever-increasing pressure of population. The 'wasteful' and 'irresponsible' use of the forests by the villagers was in turn vehemently criticized by the forest officials. It is an indisputable fact that the ever-increasing demand of the peasantry for firewood and grazing, not to speak of other forest produce, disrupted the formerly balanced relationship between agriculture and forest, and tended everywhere to jeopardize forest regeneration. This only served as a pretext for further restrictions on forest areas. The severe link between agriculture and forests at large, gave rise to an endless story of petty resistance and crime on the part of peasants against forest regulations and the local subordinates of the Forest Department.

Several newspapers commented that people experienced great oppression from the officials of the Forest Department, creating large-scale discontent in society. F. A. Lodge, Inspector General of Forests to the Hyderabad State from 1914 to 1921, wrote in his notes that 'the Forest department is generally unpopular both with other departments and with the public.' The people living in the forest zones now faced heavy fines or even imprisonment.

---

4 In fact, newspapers like *Golkonda Patrika* requested the state government to establish *forest panchayats* involving people living in nearby villages of the forest areas for the development and preservation of state forests, by giving free access to grazing and fuel needs of villagers in open forests to reduce the pressure on reserved and protected forests.

The commercialization of forests led to increase in number of timber merchants and private contractors. Traditional forest dwellers could now see these timber merchants and contractors felling trees and taking away the timber to nearby market towns, while they were not permitted to use even a fraction of the resources for their day-to-day needs. Discontent with the new forest rules and regulations manifested itself in many ways like non-compliance with imposed regulations – this was evident when peasants and tribes gave misleading information at the time of settlement of rights; 'malicious incendiarism' in the forest areas was one form of common discontent.

The peasants were traditionally dependent on forests in a number of ways. The forest was a source of firewood, manure, grass, wood for house construction and making agricultural implements, grazing space for cattle, and a great variety of minor products such as fruits, flowers, thatching material, thorns, brushwood and stakes for hedges, honey and resins, etc. In the working plans and in the reports of the Forest Department of the Hyderabad State, peasants were treated as a nuisance and an obstacle to proper forest management. All forest areas, wastelands (*beedu*, *poramboku* and *gairan*) and grazing lands were demarcated as state forests, leaving people completely deprived of their traditional grazing facilities for cattle and their customary rights to have access to fodder, fuel and wood for agricultural implements. If they were found dwelling or wandering in these forests, it was considered a serious punishable offence. Domestic animals were held up whenever they were found inside the state forests. The forest guards troubled people for trivial matters such as crossing reserved forests to reach their agricultural fields. For local communities bribing forest guards to get access to forest produce was a regular phenomenon in the Hyderabad State. If more valuable timber was required something extra money had to be paid by the individual who wanted it. The poor villagers who were being harassed by revenue and forest staff naturally developed animosity against the forest officers.

Impact of forest restrictions on people was severely felt during the occurrence of famines. India from the eighteenth century onwards had experienced a series of famines recurring at very short intervals bringing untold suffering and mortality. During famines common people used to depend on natural resources like forests for sustenance. People of Hyderabad State had traditionally evolved their own survival strategies based on their prevalent natural resource base. Forests in particular used to provide traditional support in absorbing the shock of disasters like scarcity, drought and famine. However, after the introduction of state forestry, dependence of people on forest resources was restricted and often denied completely. From the 1870s till the beginning of the twentieth century, the state suffered the worst famines and their impact was severe compared to the famines occurred earlier. The reason behind this situation was the interference of the state government with restrictions on the customary rights and traditional strategies enjoyed by the people.

It has been observed that though the cause of famines was largely due to failure of rains, severity of their impact was felt more due to the policies of the state government e.g., control over forest resources, grain exports to British India and Europe, low wages and high prices. The state government and its departments sometimes refused to acknowledge the occurrence of famine or scarcity in the region which made the bad situation to worse, as observed by the contemporary newspapers. Though some relief measures like temporary permission of the collection of forest produce, remission to or deferred payment of grazing fees, arrangements to supply fodder and the expansion of forest works, were provided, they turned out to be quite inadequate. Though during the severe famines of 1876–1878, 1899–1900 and 1918–1920, all types of forests were thrown open officially by the Nizam State for grazing, fuel and food collection, in practice the lower level staff of the Forest Department did not give full access to forest resources to the people. To a large extent inadequate relief measures by the State government led to popular discontent and large-scale deaths.

The inter-departmental rivalry between the Forest and Revenue Departments always created a rift in the official circles as each department tried to push its frontier into new areas of revenue and the forest zones were the source of those revenue areas. Even after the establishment of the Forest Department, the agricultural frontier was not closed in the State and there was a continuous expansion that took place till the 1940s. The main reason behind the agrarian expansion was population pressure (which increased from 98 lakh in 1881 to 186 lakh in 1951), frequent drought conditions and regular famines. There was also state effort to improve irrigation facilities so that the existing and the newly expanding agrarian zones could be brought under multiple cropping system. The Government with schemes such as *Ijara* and colonization encouraged bringing fresh lands under cultivation and forestlands were the most favourable tracts for expansion of arable land. The government also thought of greater revenue, which accrued through the development of irrigation. Within the State, the area of cultivated land expanded from 161.45 lakh acres to 265 lakh acres between 1890 and 1945. There was a continuous increase in cultivated area from the last decades of the nineteenth century, but significant expansion took place after 1920 due to grounding of various irrigation projects. The increase in the area of arable land was not limited to the subsistence agricultural sector alone, but also for commercial crops such as cotton, sugarcane, chilli, groundnut, etc. This expansion of agriculture was carried out mostly in the displaced large areas of forests.

Though the forest officials, throughout the period, competed for control of forest resources against the revenue officials, they could not stop the conversion of part of the forests into farmlands. In the name of promoting agriculture in forest areas, large tracts of timber producing forests were cleared without taking measures for regeneration. Considerable areas of forests were lost during the 1920s and 1930s and it is in this period that the forest area decreased from 80 lakh acres to 61.70 lakh acres due to expansion of arable lands. Thus, the earlier policy of pushing back the forests and extending the area under cultivation continued till the middle of the twentieth century. The forests were still considered by the State an obstacle to agriculture rather than a valued resource in their own right.

The forest revenue in Hyderabad State was of minor significance constituting only one per cent of the total revenue of the State. However, when considered individually, and in comparison with the other departments it occupied a very significant place in the revenue complex of the State. Moreover, there was a continuous increase of gross forest revenue from Rs. 1,09,354 in 1881 to around Rs. 20 lakh in 1930s and Rs. 1,14,05,876 in 1945. In particular, during the Second World War period the demand steeply rose to meet wartime needs of the British and this seriously affected the commercially valuable forests in the Hyderabad State.

There were few outlets for people to vent their grievances. Neither there were any conferences held in the Hyderabad State on forest grievances nor there were any official grievance committees, as in the British India. The newspapers also did not pay much attention to the grievances of the forest people. It was only after the 1930s that the press started providing vent to peasant grievances. After the 1930s, the *Golkonda Patrika* published series of articles and news items relating to the grievances of local peasants, in which forest grievances were also highlighted. Some of the contemporary newspapers like the *Deccan Post,* the *Hyderabad Chronicle,* the *Evening Mail* and the *Meezan* also reflected the mass opinion. *Golkonda Patrika* realized the importance of forests in the agrarian society like the Hyderabad State and argued for a better management of natural resources like forests. It agreed that more severe regulations might be appropriate in reserved forests, particularly in cutting commercial timber. For all other forest produce and grazing rights, particularly in open forests, it believed that there was no requirement of any prohibition or detailed system of fees and permits. It held that the greed of the Forest Department to make money through forest produce was the most important cause for all the trouble. In addition, it advised the department to limit its role to purely technical and advisory matters. The principle objection against rules and regulations was that they involved constant supervision and petty interference by the forest subordinates.

Along with the newspapers nationalist organizations also took a proactive stance in voicing people's grievances. For instance, the Andhra Maha Sabha requested the government to allow people to use forest resources for their fuel needs. They also demanded for free supply of timber for agricultural needs and asked to remove all obstacles to procure their forest needs. One of the major reasons for the famous Telangana Armed Struggle was the accumulated grievances of small peasants and landless population, and there were individual instances of conflict like that of the Babijheri case.[5]

There were many changes that occurred in the landscape and management of natural resources during the nineteenth and early twentieth centuries

---

5 The policy of clearing large tracts of forests including old, established villages inhabited for many generations, led to the only case of armed resistance by Gonds in Adilabad district in the history of Hyderabad State. It was known as Babijheri incident after the locality in which it occurred. At this place the tribals, mainly Gonds and Kolams, were subjected to severe restrictions on the practice of shifting cultivation and their movements in the reserved forest areas. In a sheer desperation under unmitigated oppression and exploitation, the helpless tribals rallied behind their leader Kumram Bhimu, in the year 1940. As most of the land in his native village passed into the hands of non-tribals, Bhimu made repeated but unsuccessful attempts to contact the state officials. After petitions for resettlements were ignored, the tribals established a settlement on their own and began to clear forests for cultivation in a locality called as Babijheri. This village was included under reserved forest in course of time and the forest guards burnt the village when tribals did not evacuate the village. Petty forest officials with the help of police constables burnt down some of the houses without any warning. One armed party shot and injured Bhimu and the followers of Bhimu were enraged and beat the forest officials. Later they took refuge in the hills. When the police asked them to surrender, they were met with the counter demand that Gonds and Kolams should be given possession of the land they had begun to cultivate. The police and forest subordinates wanted to punish tribals because they refused to surrender. They thereupon opened fire, killing Bhimu and eleven other Gonds and wounded many others. (But according to unofficial reports by police 127 Gonds were killed in that incident; The *Golkonda Patrika*, 30.9.1940; CVF. Haimendorf, *Tribal Hyderabad*, p. 122.) Even though their movement was ruthlessly suppressed, this episode succeeded in drawing the attention of the administrators and the rulers to the pathetic condition of tribals and unscrupulous exploitation by the forest subordinates, police and excise officials.

in Nizam's Hyderabad State. A large area of the forest land in the State suffered serious degradation, which can be attributed to the expansion of agriculture, relentless pressures arising from an ever increasing demand for fuel wood, fodder and timber and diversion of forest lands to non-forest uses without ensuring conservatory reforestation and essential environmental safeguards. The tendency was to look upon forests as a revenue earning source. The demand for allotment of fresh lands began to increase with the growth in rural population since the second decade of the twentieth century. The percentage of grazing lands to total area of a village that was formerly fixed at 25 per cent was brought down to 10 per cent and even 4 per cent in some cases by the 1940s. Even the areas that had considerable forestlands, like the districts of Warangal, Karimnagar, Nizamabad and Adilabad, the percentage remained just above 10 in 1935. If the forest area was excluded, the percentage of grazing area was estimated to be only 7.3 per cent for the State as a whole. This shrinkage of grazing and wastelands put more pressure on the existing forests, leading to rapid deforestation.

The worst sufferers in this fast changing forest landscape were the tribal people. Very little was given away to the them. The peasants living in the adjacent villages to the forests, who depended on forest resources, also suffered. Whenever the state was forced to concede, the concessions thus secured were insignificant to needy tribes and poor peasants. On the other hand, the government, under its colonization scheme, granted lakhs of acres in 1924 to capitalist entrepreneurs. In addition to this, a policy was made after the Second World War to allot cultivable lands to those soldiers who distinguished themselves in the war, at the rate of 60 acres each. These categories of capitalists, nobles, soldiers, etc. neither stayed in the allotted lands nor showed interest in developing them. They either allowed the lands to remain fallow, or in most cases sold them away for paltry prices to local landlords who took full advantage of the situation to further augment their land monopolization.

There were other factors that were seen as responsible for fall in the forest cover in the Hyderabad State. The effect of the deforestation probably combined with a gradual, nearly imperceptible, diminution of the rainfall over long periods. This accounted for the shrinkage of water table in the south-western, western and central portions of the Hyderabad State. The first signs of climatic change were observed during the 1930s. Contemporary observers like Captain Leonard Munn, a geologist who worked extensively in the Hyderabad State, recognized that the failure of rainfall and increase in temperature could be attributed to degradation of forests in the State. The general rise in temperature and the velocity of winds corresponded with the absence of rainfall. Occurrence of a series of famines in the second half of the nineteenth century, according to him, with more intensity than the earlier ones, was the cumulative effect of natural and man-made factors. All these famines were more than a natural swing of climatic mood. The severe famine of 1876–78, which had such catastrophic consequences, might have been due to excessive deforestation, encouraged under *Ijara* and colonization schemes. After more than five decades of haphazard felling, the ecological balance changed and the natural regenerative process no longer functioned effectively in the dry areas of Karnataka region and partly in the Telangana districts. During the 1930s, the subsoil, which was crucial for water storage, was found dried out substantially in the dry districts of the state e.g., Raichur. Many forests were ruined and several areas became treeless. Similar conditions were witnessed in other districts due to the effect of entire destruction of vegetation, which caused a serious diminution of the water table, combined with very serious erosion of the black cotton-soil upon the slopes. After deforestation, the exposed soil in the plains rapidly became saliferous. Earlier the forest cover had offered sufficient protection against this, it no longer happened. Because of the increase in water evaporation, salt moved to the surface and was deposited in the topsoil, leading to a reduction in soil's capacity to retain water. Lack of vegetation also added to high temperature during summer and the environment became highly conducive for the rapid spread of tropical diseases like malaria. Famines and epidemics indicated the change in ecological balance of the deforested regions in the State.

Extermination of wild animals in the later nineteenth and early twentieth century shows how major ecological changes could be compressed into a short span of time. The incentives of money and prizes led to the large-scale killings of wild animals that were considered to be dangerous, such as the tiger, leopard, bear, etc., by the nobles, the British and local people in the Hyderabad State. By the 1930s many of these species of animals were under the threat of extinction and government took some important measures such as setting up wildlife sanctuaries in the forest areas. This can be understood as one of the changes made by the State in the natural world in the 'remaking of forests'.

Thus, the state management of natural resources like forests marked a watershed in Hyderabad State as well. The introduction of railways and other communications at all India level made changes in the management of natural resources even in princely States like Hyderabad. It was from the second half of the nineteenth century, under the influence of the British, that the State evolved a policy of monopolist control and exploitation of forest tracts. The establishment of the Forest Department had initiated a new management of wooded zones and there were efforts to restrain the impact of overwhelming pressures (both commercial and agricultural) on forest tracts. The state takeover of large tracts of forest lands within a short span of time brought far reaching changes in the pattern of settlement, use of forest resources and agrarian extension. However, the Forest Department as a new administrative unit emerged as a new power in rural and tribal areas. The edifice of formal forestry in the Hyderabad State remained a major legacy of the British colonial policy. Like in the British ruled areas, the State monopolistic control over forests had brought severe miseries in the lives of tribes and peasants that were directly or indirectly associated with access to forest resources.

Appendix

Appendix I. Famines and its causes in Hyderabad State

Year	Area affected	Causes
1676–77	Gulbarga district	Apparently caused in the first place by excessive rains
1685	Western part of the state	War prevented any relief
1702–4	Western part of the state	Excessive rainfall and war prevented relief measures
1804	Lingsugur, Raichur, Nagar-Kurnool, Aurangabad, Hyderabad, Bid, Bidar and Parbhani districts.	Failure of rains leading to drought; kharif and rabi crops failed. Except in Aurangabad where effects of famine were coupled with Holkar depredations
1813	Shorapur	Owing to pressure of prices
1819	Gulbarga, Bid, Lingsugur, Indur, Parbhani, Shorapur	Excessive and incessant downpours
1825	Parbhani and Aurangabad	Excessive downpour
1833	Gulbarga, Lingsugur, Nagar-Kurnool, Indur, Hyderabad and Shorapur	Failure of monsoon
1846	Western portion of the state	Rise in price of grains
1854	Gulbarga, Shorapur, Raichur, Nagar-Kurnool, Indur and Hyderabad	Drought
1862	Hyderabad	Insufficient rainfall leading to crop failure

(contd.)

(contd.)

Year	Area affected	Causes
1866	Hyderabad and it neighbourhood	Insufficient rainfall
1871	Aurangabad, Nagar-Kurnool and Indur	Season of drought; kharif crops failed but rabi crops were good in Indur and Nagar-Kurnool
1876–77 1877–78	Great portion of Lingsugur, half of East Raichur, 1/3 of Shorapur, 1/4 of Gulbarga, 1/8 of Bid and Naldurg, besides Nalgonda and Nagar-Kurnool witnessed scarcity	Failure of rain
1890	Western part of the State	Failure of rain
1894	Mainly western part of the State	Failure of rain
1897	Raichur, Lingsugur, parts of Gulbarga, Osmanabad and Bid	Failure of rain
1898	Western portion of the state	Failure of rains in 1899, but grain was enough for local needs
1899–1900	Mainly western parts of the state	The great famine of India affected the state also; export of grain to adjoining districts of British India leading to high prices
1906–12	Mainly western part of the state	Scarcity conditions
1913	Marathwada fodder famine	Failure of rains
1918–19; 1919–20	Western border i.e., Aurangabad, Bir, Warangal, Karimnagar, Mahaboob-nagar Adilabad, Medak, Nalgonda	Excessive and untimely rains plus rise in prices due to great European War; export of grains
1927	Raichur and parts of Nalgonda district	Scarcity conditions
1934–35	Parts of Raichur, Gulbarga and Karimnagar districts	Scarcity conditions
1940	Whole of Telangana and partly Marathwada and Karnataka	Scarcity conditions

Source: Maulavi Syed Mahadi Ali, *Report on the History of the Famine in His Highness the Nizam's Dominions in 1876–77, 1877–78*, Exchange Press, Bombay, 1879; HEH the Nizam's government, Revenue Department, Hyderabad, *Scarcity and Famine Manual*, Government Central Press, Hyderabad, 1938; *Karimnagar District Gazetteer*, (1972), pp. 102–3 and A. Loveday, *The History and Economics of Indian Famine*, pp. 135–8.

Appendix II. Land utilization abstract of Hyderabad State for the year 1953–54 (in acres)

Districts	Geographical area	Forest	Barren and uncultivated land	Land put to non-agri. uses	Cultivable waste	Permanent pastures	Misc. trees	Current fallows	Other fallow land	Net area sown
Adilabad	4157494	1098595	658266	223179	160759	159160	23757	202250	163353	1468175
Aurangabad	8992090	174154	51988	123293	47612	254751	644	123608	257310	2958530
Bidar	2977373	29606	76452	88461	94247	144717	24158	330708	340259	1848770
Gulbarga	2721506	25883	21065	61045	49961	123809	3291	274914	277909	1883629
Hyderabad	4504869	95868	225955	121887	101154	150093	15971	224274	167945	3401716
Karimnagar	1052375	93378	109553	69952	139072	113788	12507	150250	89013	374862
Khammam	2957749	638830	215542	170712	77587	153474	26108	347931	67852	1259713
Bhir	2789758	1030666	226262	151513	178869	167459	17800	103719	120255	793215
Mahaboobnagar	3869258	790746	280873	142606	86597	167551	21032	431192	129536	1819220
Medak	2175325	191256	220548	156796	86114	171258	22805	231868	62590	1032095
Nalgonda	8415672	289391	256847	148424	84590	289070	14826	396498	219039	1766987
Nanded	2488621	79868	69974	75062	32301	177663	8797	177783	163784	1703394
Nizamabad	1768972	276001	233608	142611	86481	120767	16844	136468	92140	664052
Osmanabad	2356973	–	32205	42684	39689	56280	231	253147	211566	1721171
Parbhani	3071470	42149	41694	59867	37131	119231	9936	423431	122044	2215987
Raichur	4054663	29257	240817	141723	65657	135349	12889	309892	34887	3084192
Warangal	2476588	274675	247216	143375	111884	256719	22403	217908	124495	1077913
Hyderabad State	50830845	5110318	3208865	2063190	1379705	2761139	254199	4335831	2643977	29073621
	(100%)	(10.05%)	(6.31%)	(4.05%)	(2.71%)	(5.43%)	(0.50%)	(8.52%)	(5.20%)	(57.19%)

Source: *Questionnaire of the Forest Policy Committee,* Hyderabad State, Hyderabad, 1956, p. 14.

Bibliography

Primary Sources

Unpublished original sources

Andhra Pradesh State Archives, Hyderabad

HEH the Nizam's Government, Office Files of the Forest Department; Revenue Department; Agriculture Department; Rural Reconstruction Department; Judicial Department; Police Department; General Administration Department; Public Works Department.

Interim Repository (Andhra Pradesh Secretariat), Hyderabad

HEH the Nizam's Government and Government of Hyderabad, Office Files of the Forest Department; Revenue Department and Finance Department (Supply and Resource Branches).

Board of Revenue, Hyderabad

HEH the Nizam's Government and Government of Hyderabad, Office Files of the Forest Department; Revenue Department and District Administration Files.

National Archives of India, New Delhi

Hyderabad Residency Files and Crown Representative Reports (Micro Films).

All India State Peoples' Conference, File No. 65, Part 1, Nehru Memorial Museum and Library, New Delhi.

Published original sources

HEH the Nizam's Government, *Annual General Administrative Reports – from 1884–1948,* Hyderabad.

________, *Memorandum on the Progress of Hyderabad Administration under His Excellency Sir Asman Jah, Minister – 1887–88.*

________, *Census of India*, 1891–1951, Hyderabad State Reports.

________, *Annual Administrative Reports of the Forest Department*, 1893–1948.

________, *Annual Administrative Reports of the Agriculture Department.*

________, *Report on the History of the Famine in His Highness the Nizam's Dominions in 1876–77, 1877–78*, The Exchange Press, Bombay, 1879.

________, *Report on the operations of HEH Public Works and Allied Departments for the 17 years from 1867 to 1884*, Hyderabad, 1885.

________, *Report on the Famine Relief Operations in HEH the Nizam's Dominions during 1899–1900*, Government Central Press, Hyderabad, 1901.

________, *Statistical Year Books* and *Statistical Abstracts*, Hyderabad State Reports.

________, *Notes Regarding Administration of His Highness the Nizam's the Forest Department, From 1867 to 1911 AD,* Government Central Press, Hyderabad, 1913.

________, *The Reserved State Forests Rules, 1303 Fasli,* (1893–94), published by the Revenue Department, Hyderabad, 1895.

________, *Revenue Administration Report, 1324 F (1914–1915)*, Government Central Press, Hyderabad, 1916.

________, *Souvenir of His Royal Highness the Prince of Wales' Visit, Hyderabad State*, Government Central Press, Hyderabad, 1922.

________, *Decennial Report on the Administration of the HEH the Nizam's Dominions, 1912–1922*, Government Central Press, Hyderabad, 1930.

________, *Report on the Working of the Departments Under the Director General of Revenue, 1325 F (1915–16), 1326 F (1916–17) and 1328 F (1918–19)* Central Jail Press, Hyderabad, 1917–1920.

________, *Hyderabad State, List of Officials, Nobles and Personages*, Residency Government Press, Hyderabad, 1925.

________, *Forest Code for 1335 Fasli (1925–26) for the HEH the Nizam's Dominions*, Government Press, Hyderabad, 1927.

________, Hyderabad State District Gazetteers, All districts 1931–1936.

________, *Report of the Paigah Commission of 1345 H (1926)*, Government Central Press, Hyderabad, 1937.

________, *Some Economic Facts and Figures of HEH the Nizam's Dominions*, Bulletin No. 1, Department of Statistics, Government Central Press, Hyderabad, 1937.

________, *The Economic Life of Hyderabad,* Government Central Press, Hyderabad-Deccan, 1937.

________, *Hyderabad Scarcity and Famine Manual*, Revenue Department, Government Central Press, Hyderabad, 1938.

________, *Preliminary note on soil question in certain irrigated lands in Hyderabad,* Government Central Press, Hyderabad, 1939.

________, *Working Plan Report of fuel and fodder reserves of the Western Circle of the HEH the Nizam's Dominions for 1928–1929*, Government Central Press, Hyderabad, 1930.

________, *Estimates of Area and Yield of Principal Crops in Hyderabad State, 1925–26 to 1934–35.*

________, *Engineer's Note on Nizam Sagar Project*, vol-II, Government Central press, Hyderabad, 1941.

________, *Agricultural, Industrial and Trade Statistics for the year 1343 Fasli (1933–34).*

________, *Report of the Administration of the Court of Wards Department, 1349 F (1939–40)*, Government Central Press, Hyderabad, 1942.

________, *Some Aspects of Hyderabad,* Information Bureau, Hyderabad, 1941.

________, *A Summary of Hyderabad's War Effort, 1939–44*, Government Central Press, Hyderabad, 1944.

Government of Hyderabad, *Bulletin on Economic Affairs,* vol. II, no. 2, February 1949.

________, *Report of the Agrarian Reforms Committee,* Hyderabad, 1949, Part I.

________, *Report of the Tenancy Committee*, Government Central Press, Hyderabad, 1949.

________, *Report of the Administration of Hyderabad State, Sep. 1948 – Mar. 1950*, Government Central Press, Hyderabad.

________, *A Review on Hyderabad Finance*, Finance Department, Government Press, Hyderabad, 1952.

________, *Jagir Administration,* vol. 1, Office of the Jagir Administrator, Hyderabad, 1952.

________, *A Note on the Agricultural Marketing in Hyderabad State,* Hyderabad, 1952.

________, *The Hyderabad Survey and Settlement Manual*, 1952.

________, *Questionnaire of the Forest Policy Committee*, Hyderabad with forward and explanatory notes issued by B. B. Wadhwani, Secretary, Forest Policy Committee, Hyderabad 1956.

________, *Hyderabad, 1952–1956*, Citizen Press, Secunderabad, (n.d.).

Government of India, *Famine Commission Report,* Part I, 1880.

________, *Selection of Papers Relating to the Famine of 1896–97,* vol. 5., Superintendent of Government Press, Calcutta, 1898.

________, *Imperial Gazetteer of India, Provincial Series, Hyderabad State,* Superintendent of Government printing, Calcutta, 1909, reprint, Atlantic Publications and distributors, Delhi, 1991.

________, *Indian Forest Code (General Regulations),* Calcutta, 1913.

________, A Manual of Forest Law compiled for the use of the Students at the Imperial Forest College, Dehradun, Government Press, Madras, 1913.

________, *Explanatory Notes on Forest Law,* Imperial Forest College, Dehradun, Government Press, Calcutta, 1926.

________, *Report of the Royal Commission on Agriculture in India*, Bombay, 1928, (Reprint 1983).

________, *Proceedings of the Forestry Conference,* Held on the 9 and 10 December, 1937, New Delhi, 1938.

________, Ministry of Agriculture, 'India's Forests and the War' Delhi, 1948.

________, *National Forest Policy of India,* New Delhi, 1952.

________, *History of Indian Railways*, Delhi, 1964.

________, *Forestry in India,* Ministry of Agriculture, New Delhi, 1982.

Government of Andhra Pradesh, *Report of the Land Revenue Reforms Committee, 1958–59*, Hyderabad, 1959.

________, *Report on the Economic Survey of Tungabhadra Project Area*, Hyderabad, 1959.

________, *The Andhra Pradesh Code: Law Department, Acts and Regulations in Force in the Telangana Area of the State of Andhra Pradesh 1305 Fasli – 1359 Fasli,* Vol. III, Hyderabad, 1964.

________, Forest Department, *Facts and Figures, 1994*, The Principal Chief Conservator of Forests, Andhra Pradesh, Hyderabad, 1994.

________, Andhra Pradesh District Gazetteers – Nalgonda, Warangal, Nizamabad, Adilabad, Khammam, Karimnagar, Mahabubnagar.

Government of Karnataka, Mysore District Gazetteers, Bijapur, Berar, Bidar and Gulbarga districts, 1966–1975.

Contemporary Newspapers and Journals

The Golkonda Patrika (Telugu) (Daily, Bi-Weekly and Weekly)
Meezan (Telugu, Urdu and English) (Daily)
Hyderabad Information (Monthly)
The Hyderabad Bulletin (Monthly)
The Hyderabad Forest Magazine (Quarterly)
The Deccan Geographer
The Hyderabad Government Bulletin on Economic Affairs (Monthly)
The Deccan Chronicle (Daily)
The Deccan Post (Daily)
The Hyderabad Chronicle (Daily)
The Evening Mail (Daily)
Nizam Vijay (Marathi Weekly)

Secondary Sources

Agriculturists' Association, *Immediate Agrarian Problems, Hyderabad State,* Hyderabad, 1949.

Ahmed, L. S. Habib, *An Economic Survey of the Sirpur Paper Mills Limited,* Unpublished MA (Economics) dissertation, Osmania University, Hyderabad, 1960.

Aitchison, C. U., *A Collection of Treaties, Engagements and Sanads*, vol. 9, Superintendent of government printing, Calcutta, 1909.

Ali, Abid Hassan, Syed, *Whither Hyderabad*, the B. N. Press, Madras, 1935.

Ali, Mahdi, *Hyderabad Affairs*, 10 vols., The Times of India Steam Press, Bombay, 1883–89.

________ , *Report on the History of the Famine in His Highness the Nizam's Dominions in 1876–77, 1877–78*, The Exchange Press, Bombay, 1879.

________ , *Replies to the Questions Contained in Chapter 1, etc., of the Memorandum of the Famine Commission*, Times of India Press, Bombay, 1879.

Ali, Moulvi Cheragh, *Hyderabad (Deccan) under Sir Salar Jung I*, 2 vols., Education Society's Press, Bombay, 1884.

Allen, Roscoe, P., *Memorandum on Irrigation in Hyderabad,* Government Press, Hyderabad, 1902.

Anonymous, *Social Service among the Tribes and Backward Classes in Hyderabad*, Government of Hyderabad, 1956.

Arnold, David and R. Guha (eds.), *Nature, Culture, Imperialism: Essays on the Environmental History of South Asia*, Oxford University Press, New Delhi, 1995.

Arnold, David, 'Rebellious Hillmen: The Gudem-Rampa Risings, 1893–1924', in: Ranajit Guha (ed.), *Subaltern Studies*, I, Oxford University Press, New Delhi, 1982.

________ , 'Famine in peasant consciousness and peasant action: Madras 1876–78', in: Ranajit Guha, (ed.) *Subaltern Studies*, III, Delhi, 1984.

________ , *Famines, Social Crisis and Historical Change*, London, 1989.

Ribbentrop, B., *Forestry in British India*, Government Press, Calcutta, 1900.

Baden Powell B. H., *The Land System of British India*, Vol. III, Oxford University Press, Oxford, 1892.

________ , 'On the defects of existing Forest Law (Act VII of 1865) and proposals for a new Forest Act', in: B. H. Baden-Powell and J. S. Gamble, (eds.), *Report of the Proceedings of the Forest Conference 1873-74*', Calcutta, 1874;

________ , *Forest Law*, London, 1893.

________ , *Administration of Land Records and Tenure in British India*, New Delhi, 1908. *Memorandum on Forest Settlements in India,* Calcutta, 1891.

Baig, M. M. Ali, 'Cattle improvement in Hyderabad', *The Hyderabad Forest Magazine*, (Quarterly), No. II, July, 1942, pp. 5–7.

Baker, D. E. U ., "A Serious Time": Forest Satyagraha in Madhya Pradesh, 1930', *IESHR*, Vol. 21, No. 1, 1984, pp. 71–90.

Basava Punnaiah, S. M., *Telangana Armed Struggle – Realities*, Vijayawada, 1972.

Bawa, V. K., 'Some problems in the study of Hyderabad State' in *Itihas*, Vol. I, No. 1, 1973.

________, *The Nizam Between Mughals and British – Hyderabad Under Sir Salar Jung 1*, S. Chand and Company Ltd., New Delhi, 1986.

________, *The Last Nizam: The Life and Times of Mir Osman Ali Khan*, Viking, Penguin, New Delhi, 1992.

Bharucha, S. M., *Agricultural Indebtedness in HEH the Nizam's Dominions*, Hyderabad, 1937.

Bhatia, B. M., *Famines in India,* Asia Publishing House, Bombay, 1963.

Bhattacharya, Neeladri, 'Pastoralists in colonial world', in: David Arnold and R. Guha, (eds.), *Nature Culture Imperialism,* Oxford University Press, New Delhi, 1995.

Bilgrami, S. H. and C. Willmott, *Historical and Descriptive Sketch of HH the Nizam's Dominions*, The Times of India Press, Bombay, 1883.

Bilgrami, S. H., *A Memoir of Sir Salar Jung 1*, The Times of India Steam Press, Bombay, 1883.

Brandis, D., *Forestry in India: Origins and Early Development*, Brigton,1872, Reprint by WWF India, Nataraj Publications, Dehradun, 1994.

________, *Indian Forestry*, Oriental University Institute, Woking, 1897.

________, *Suggestions Regarding Forest Administration in the Madras Presidency,* Madras, 1883.

Buchanan, Francis., *A Journey from Madras through the Countries of Mysore, Canara and Malabar*, Higginbothams and Co., Madras, 1870, 2 vols.

Champion, H. G. and S. K. Seth., *A Revised Survey of the Forest Types of India*, Government of India, New Delhi, 1968.

Chaturvedi, B. N. and K. S. Reddy, 'Irrigation in Telangana sources and problems' in: *Deccan Geographer*, Hyderabad, January 1964.

Chaudhuri, Nani Gopal, *British Relations with Hyderabad: 1798–1843*, University of Calcutta, Calcutta, 1964.

Claude, C. A., *Glimpses of the Nizam's Dominions*, Bombay and London, 1898.

Crosby, Alfred W., *Ecological Imperialism: The Biological Expansion of Europe, 900–1900*, Cambridge University Press, 1986.

Damodaran, A., 'Tragedy of the commons and comedy of common property resources', *EPW,* Sept. 21, 1991, pp. 2213–2215.

Damodaran, Vinita., 'Famine in a forest tract: Ecological change and the causes of the 1897 famine in Chota Nagpur, Northern India', in: Richard Grove, Vinita Damodaran and Satpal Sangwan (eds.), *Nature and the Orient,* pp. 853–890.

________, *Broken Promises: Popular Protest, Indian Nationalism and the Congress Party in Bihar, 1935–1946*, Oxford University Press, New Delhi, 1992.

Dangwal, Dhirendra Datt., 'Forests, farms and peasants: Agrarian economy and ecological change in the U.P. hills, 1815–1947', *Studies in History*, vol. 14, no. 2, July–December, 1998, pp. 349–371.

________, 'State, forests and graziers in the hills of Uttar Pradesh: Impact of colonial forestry on peasants, Gujars and Bhotiyas', *IESHR*, vol. 34, no. 4, 1997, pp. 405–435.

Das, Dipakranjan, *Economic History of the Deccan: From the First to the Sixth Century AD*, Munshiram Manoharlal Publications, Delhi, 1967.

Department of Information and Public Relations, *History and Legend in Hyderabad*, Osmania University Press, Hyderabad, 1953.

Desai, A. R., (ed), *Peasants Struggle in India*, Oxford University Press, Bombay, 1979.

Dhanagare, D. N., 'Social origins of the peasant insurrection in Telangana', *Contributions to Indian Sociology*, Vikas Publishing House, New Delhi, 1974.

________, *Peasant Movements in India 1920–50*, Oxford University Press, Delhi, 1983.

Dube, S. C., *Economic Life of Hyderabad,* Government Central Press, Hyderabad, 1937.

Dutt, Romesh, *Indian Famines: Their Causes and Prevention*, P. S. King and Sons, London, 1901.

Elwin, V., *Leaves from the Jungle*, London, 1958.

________, *The Tribal World of Verrier Elwin: An Autobiography*, Delhi, 1964.

Forest Research Institute, *100 Years of Indian Forestry, 1861–1961*, 2 vols., Dehradun, 1961.

Gadgil M. and Ramachandra Guha, 'Greening the Commons', *Mainstream*, January 2, 1989, pp. 9–10 and 30.

________, *This Fissured Land: An Ecological History of India*, Oxford University Press, New Delhi, 1992.

Gadgil, M. and V. D. Vartak, 'Groves dedicated to God', *Illustrated Weekly of India*, Vol. 35, 1973, pp.19–21.

Gadgil, M. and V. D. Vartak., 'Sacred groves of India: A plea for continued conservation', *Journal of Bombay Natural History Society*, vol. 72, 1975, pp. 314–330.

Gadgil, Madhav and F. Berks, *Traditional Resource Management Systems: Resource Management and Optimisation*, 1991.

Gadgil, Madhav and P. Iyer., 'On the diversification of common-property resource use by Indian Society', in Fikret Berks (eds.), *Common Property Resources: Ecology and Community-based Sustainable Development,* Belhaven Press, London, 1989.

Gadgil, Madhav and R. Guha., *This Fissured Land: An Ecological History of India*, Oxford University Press, New Delhi, 1992.

Gadgil, Madhav, 'Forest management, deforestation and people's impoverishment', *Social Action*, vol. 39, Oct–Dec, 1989, pp. 357–383.

________, 'The Indian heritage of a conservation ethic', in: B. Allchin, E. R. Allchin and B. K. Thapar (eds.), *Conservation of the Indian* Heritage, Cosmo Publications, New Delhi, 1989.

Gadgil, M., S. N. Prasad and Rauf Ali., 'Forest management and forest policy in India: A critical review', *Social Action*, vol. 33, no. 4, April–June, 1983, pp. 127–155.

Gafoor, Syed Khaja Abdul., *Tribes and Tribal Welfare in Hyderabad,* Government Press, Hyderabad, 1953.

Gopalakrishnaiah, Vavilala., 'Nizam Jagirula Raddu Shasanam Lopalu' (Telugu), *Andhra Prabha* (Daily), 18 September, 1949.

Gopalan, M., 'Scheme for the agricultural development under the irrigation projects in Hyderabad', *Hyderabad Government Bulletin on Economic Affairs*, vol. 1, no. 9, December, 1948.

Gribble, J. D. B., *A History of the Deccan*, vol. I and II, Luzac & Co., London, 1896.

Grove, R. H., 'Conserving eden: The (European) East India companies and their environmental policies on St. Helena, Mauritius and in Western India, 1660–1854', *Comparative Studies in Society and History*, vol. 35, no. 2, April 1993, pp. 318–351.

________, *Green Imperialism: Conservation and Colonial Expansion, 1600–1860*, Cambridge University Press, Cambridge, 1995.

Grove, R. H., V. Damodaran and S. Sangwan (eds.), *Nature and the Orient: The Environmental History of South and Southeast Asia*, Oxford University Press, New Delhi, 1998.

Grove, R. H. and D. Anderson (eds.), *Conservation in Africa: People, Politics and Practice*, Cambridge, 1987.

Guha, Ramachandra., 'Forestry in British and Post-British India: A historical analysis', in: *EPW*, vol. 18, no. 44 and 46, Oct–Nov., 1985, pp. 1882–1896 and 1940–1946.

________, 'Pre-history of Indian Environmentalism: Intellectual tradition', *EPW*, vol. 20, nos. 45–7, November, 1985, pp. 1939–52.

________, *The Unquiet Woods: Ecological Change and Peasant Resistance in the* Himalaya, Oxford University Press, New Delhi, 1989.

________, 'An early environmental debate: The making of the 1878 Forest Act', *IESHR*, vol. 27, no. 1, 1990, pp. 65–84.

________, 'Writing Environmental History in India', *Studies in History*, vol. 9, no.1, Jan–June, 1993, pp. 119–150.

________, 'Social-ecological research in India: A 'status' report', *EPW*, February, 1997, pp. 345–352.

Guha, R. and M. Gadgil, 'State forestry and social conflict in British India: A study in the ecological bases of agrarian protest', *Past and Present,* vol. 123, May, 1989, pp. 141–177.

Guha, Sumit., 'Forest polities and agrarian empires: The Khandesh Bhills, c. 1700–1850', *IESHR*, vol. 33, no. 2, 1996, pp. 133–153.

Gupta, Akhil., 'Revolution in Telangana, 1946–51', *Social Science Probings*, vol. 3. no. 1, March 1986.

Gupta, G. S., 'Gonds and their agony', *The Golkonda Patrika*, 3.7.1941.

Habib, Irfan and Raychaudhuri (eds.), *The Cambridge Economic History of India, c. 1200–1750,* vol. 1. New Delhi, 1982.

Haimendorf, CVF., 'Tribal populations of Hyderabad yesterday and today' an essay in 1941 Census, HEH the Nizam's Dominions, Part I, (Appendix I), pp. XXIV–XXV.

________, *The Aboriginal Tribes of Hyderabad*, 3 vols., London, 1943–1948.

________, *Tribal Hyderabad – Four Reports*, HEH the Nizam's Government, Government Central Press, Hyderabad, 1945.

________, *The Gonds of Andhra Pradesh: Tradition and Change in an Indian Tribe*, London/Delhi, 1979.

________, *Tribes of India: The Struggle for Survival*, Oxford University Press, New Delhi, 1982, (reprint 1994).

________, *Life Among Indian Tribes: The Autobiography of an Anthropologist*, Oxford University Press, New Delhi, 1990.

Hassan, Syed Abid, *Whither Hyderabad* (no details).

Iyangar, S. Kesava., *Rural Economic Enquiries in the Hyderabad State 1949–51,* Hyderabad State, government press, Hyderabad, 1951.

________, *Economic Investigations in Hyderabad State – 1929–30*, 3 vols., HEH the Nizam's government, Government Central Press, Hyderabad, 1931–32.

________, *A Review of Hyderabad Finance*, (n.d.).

Joshi, Gopa., 'Forests and forest policy in India', *Social Scientist,* vol. 11, no. 1, Jan, 1983, pp. 43–52.

________, 'Forest policy and tribal development: Problems of implementa-tions, ecology and exploitation', *Social Action*, vol. 31, Oct–Dec. 1981, pp. 446–468.

Kate, P. V., *Marathwada Under the Nizams*, Mittal Publications, Delhi, 1987.

Khan, Abdul Quayyaum., *Irrigation Hyderabad State,* in Hyderabad Government bulletin in Economic Affairs, vol. I, no. 9, December, 1948.

Khan, Fathaulla., *A History of Administrative Reforms in Hyderabad State*, New Hyderabad Press, Secunderabad, 1935.

Khan, Mohammed Sharfuddin., *Forest Flora of Hyderabad State,* Government of Hyderabad, Government Press, Hyderabad, 1953.

Khan, Raza Ali., *Hyderabad 400 Years (1591–1991)*, Zenith Services, Hyderabad, 1991.

Khan, S. H. Hussain., 'Forest Policy for the dominions and its necessity', In: *Proceedings of The Hyderabad Forest Conference, 1332 Fasli, (1923),* 1924, Government Press, Hyderabad.

Khan, Sadath Ali., *Projects for Prosperity*, Government Central Press, Hyderabad, 1951.

Khan, Yousuf Hussain., *Nizam-ul-Mulk Asaf Jah 1, Founder of Hyderabad State*, Basel Mission Press, Mangalore, 1935.

Khusro, A. M., *Economic and Social Effects of Jagirdari Abolition and Land Reforms in Hyderabad*, Osmania University Press, Hyderabad, 1958.

Klein, Ira., 'When the rains failed: Famine, relief and mortality in British India', *IESHR*, vol. 21, no. 2, 1984, pp. 185–214.

Kumar, A. Biju., 'Sacred groves: The virgin forests', *Science Reporter*, vol. 35, no. 10, October, 1999, pp. 10–13.

Kumar, D., 'The evolution of colonial science: Natural history and the East India Company', in: J. M., MacKenzie, (ed.), *Imperialism and the Natural World*, Manchester, 1990, pp. 51–66.

Kumar, Deepak., (ed.), *Science and Empire: Essays in Indian Context (1700–1947)*, Anamika Prakashan, NISTADS, Delhi, 1991.

Law, John., *Modern Hyderabad (Deccan)*, Calcutta, 1914.

Lodge, F. A., *Notes on Forest Policy and Forest Management*, Bulletin Press, Hyderabad, 1916.

Loveday, A., *The History and Economics of Indian Famines,* G. Bell and Sons Limited, London, 1914, Reprint, Usha Publications, New Delhi, 1985.

Ludden, David., 'Ecological zones and the cultural economy of irrigation in South Tamilnadu', *South Asia*, vol. 1, no. 1, 1978, pp. 1–13.

________, *Peasant History in South India*, Princeton, 1984.

Lynton, H. R., and M. Rajan., *The Days of the Beloved*, Orient Longman, Hyderabad, 1987.

MacKenzie, J. M., (ed.), *Imperialism and the Natural World*, Manchester, 1990.

Madhavan, P. Setu., *18th Century Deccan* (no date).

Mason, L., *Note on the Progress made by the Hyderabad Forest Department since My Report as Forest Advisor in 1341 Fasli*, Government Central Press, Hyderabad, 1938.

Mc Aucliff, R. P., *The Nizam, the Origin and Future of the Hyderabad State*, Cambridge University Press, Cambridge, 1904.

Mirza, Khurshid., *A Brief Outline of the Geological History of Hyderabad with Reference to its Mineral Resources*, Hyderabad, 1937.

Mohan, Ram., 'The Telangana peasant armed struggle, 1946–51', *EPW*, vol. 8, no. 23, 1973.

Morris D. Morris., 'What is a famine?' *EPW*, IX, 2, Nov. 1974, pp. 1855–1864.

Mudiraj, K. K., *Pictorial Hyderabad*, Vol. II, Chandrakanth Press, Hyderabad, 1934.

Munn, Capt. Leonard., 'Geology of the underground water resources of the Hyderabad State and notes on well sinking', *Journal of Hyderabad Geological Society: Water Supply*, vol. II, part 2, Government Central Press, Hyderabad, 1934.

Munshi, K. M., *The End of an Era: Memories*, Bharatiya Vidya Bhavan, Bombay, 1957.

Murali, Atlury., 'Alluri Sitarama Raju and the Manyam rebellion of 1922–24', *Social Scientist*, Vol. 131, April 1984.

________, 'Civil Disobedience Movement in Andhra 1920–1922: The nature of peasant protest and the methods of Congress political mobili-sation', Kapil Kumar (ed.), *Congress and Classes: Nationalism, Workers and Peasants,* Manohar Publishers, New Delhi, 1988.

________, 'Whose trees? Forest practices and local communities in Andhra, 1600–1922', David Arnold and R. Guha, (eds.), *Nature Culture Imperialism*, Oxford University Press, New Delhi, 1995.

Narayan, B. K., *Agricultural Development in Hyderabad State, 1900–1956*, Keshav Prakashan, Secunderabad, 1961. *A Survey of Industries in Telangana*, Secunderabad, 1962. *Finances and Fiscal of Hyderabad State (1900–1956)*, Government of Andhra Pradesh, Hyderabad, 1973.

Nayeem, M. A., *The Splendour of Hyderabad*, Jaico Publishing House, Bombay, 1987. *Mughal Administration of Deccan under Nizam ul Mulk Asaf Jah (1720–48)*, Jaico Publishers, Bombay, 1985.

Odum, Eugene P., *Fundamentals of Ecology*, W. B. Sanders Company, Philadelphia, 1971.

Osmania Graduates' Association., *Development of Hyderabad (The Part Played by Engineers)*, Hyderabad, (n.d).

Palmer, George., *Report on the Operations of Public Work and Allied Department, 1844–45*. H. H. Nizam's Government, , P. W. D. Press, Hyderabad, (no date).

Patridge, E. A., *Forest Flora of Hyderabad State,* H. E. H. the Nizam's Government, Government Central Press, Hyderabad, 1911.

Pavier, Barry., 'The Telangana Armed Struggle', In: *EPW*, Special number, August 1974, vol. IX, p. 1413. *The Telangana Movement: 1944–51*, Vikas Publishing House, New Delhi, 1981.

Poffenberger, M. and McGean (eds.), *Village Voices, Forest Choices: Joint Forest Management in India*, Oxford University Press, New Delhi, 1996.

Pouchepadass, Jacques., 'The ecological history of the central Western Ghats in the modern period: Preliminary survey', in: *Journal of Indian Anthro-pological Society*, vol. 26, no. 1 and 2, March–July, 1991, pp. 121–138.

________, 'British attitudes to shifting cultivation in colonial South India: A case study of Canara District, 1800–1920', in: David Arnold and R. Guha (eds.), *Nature Culture Imperialism,* Oxford University Press, New Delhi, 1995.

Prasad, S. N., *Operation Polo: The Police Action Against Hyderabad*, Government of India, Delhi, 1972.

Pullaih, T., and B. R. Prasad Rao., *Flora of Nizamabad (Andhra Pradesh),* Bishen Singh Mahendra Pal Singh, Dehradun, 1995.

Qureshi, A. I., *The Economic Development of Hyderabad*, vol. I, Orient Longman Ltd., Madras, 1941.

________, 'Land tenures in Hyderabad – A historical survey', *Indian Journal of Economics*, vol. XXVII, 1946–1947.

Raj, Sheela., *Medievalism to Modernism*, Popular Prakashan, Bombay, 1987.

Raju, S. P., *History of the Nizam Sagar Project, Hyderabad– Deccan*, Government Central Press, Hyderabad, 1939.

Ram, N., 'Impact of early colonization on economy of South India', *Social Scientist*, February 1973.

Raman, K. N. Anantha., *The Report of the Land Revenue Reforms Committee, 1958–59*, Government of Andhra Pradesh, Hyderabad, 1959.

Ramesan, N. (ed.), *The Freedom Struggle in Hyderabad,* vol. IV, Hyderabad, 1966.

Ranga, N. G., *Economic Organisation of Indian Villages*, vol. II, Bombay, 1929.

Rangarajan, Mahesh., *Fencing the Forest: Conservation and Ecological Change in India's Central Provinces, 1860–1914*, Oxford University Press, New Delhi, 1996.

________ , 'Imperial agendas and India's forests: The early history of Indian Forestry', in: *IESHR*, vol. 31, no. 2, pp. 147–167.

Rao, Ch. Hanumantha., *Agricultural Growth, Rural Poverty and Environmental Degradation in India,* Oxford University Press, New Delhi, 1994.

Rao, Ch. Rajeshwara., Arutla Ramachandra Reddy and Y. V. Krishna Rao., *Bhooswamya Vidhanam Raddukai Raithula Veerochita Tirugubatu*, Telugu., (Heroic Peasants' Struggle against the Landlord System) Andhra Pradesh Communist Samithi, Vijayawada, 1973.

Rao, CVL. Narsimha., 'Details of the Forest Department in the Nizam's Dominions' (Telugu), *Golkonda Patrika,* 9.5.1938.

Rao, D. Venkateshwara., *The History of the Telangana People's Armed Struggle (1946–51)*, Proletarian Line Publications, Hyderabad, 1988.

Rao, D. V., *Manual on Groundwater and Well-Sinking*, Hyderabad, 1947.

Rao, K. Mohan., *The Kolams: A Primitive Tribe in Transition*, Booklinks Corporation, Hyderabad, 1990.

Rao, M. Narasinga., *50 Samvatsaramula Hyderabad*, (Telugu), M. N. Smaraka Samithi, Hyderabad, 1977.

Rao, Neena Ambre., *Forest Ecology in India: Colonial Maharashtra 1850–1950,* Cambridge University Press India Pvt. Ltd., New Delhi, 2007.

Rao, Raghunadha Gangaraju., 'Village problems in Nizam's State', in: *The Golkonda Patrika*, 5.4.1934.

Rao, R. V., *A Review of Hyderabad Finance*, Hyderabad, 1951.

Rao, V. Manikya., *Hyderabad Swatantrodyama Charitra* (Telugu), Swatantra Samarayodhula Sangham, Hyderabad, 1984.

Rawat, Ajay., *Indian Forestry: A Perspective*, Indus Publishing Company, New Delhi, 1993.

Ray, Bharati., *Hyderabad under British Paramountcy*, Oxford University Press, New Delhi, 1988.

Ray, P. C., *Indian Famines: Their Causes and Remedies*, Cherry Press, Calcutta, 1901.

Reddy, A. Ramachandra., *Telangana Porata Smrutulu*, Visalandhra Publishing House, Hyderabad, 1988.

Reddy, M. P. R., *Peasants and State in Modern Andhra History*, Clio Book Club, Kavali, 1986.

Reddy, Ravi Narayana., *Heroic Telangana: Reminiscences and Experiences*, Communist Party of India, Delhi, 1973.

Reddy, Suravam Pratapa., *Pradhamika Swatyamulu*, Hyderabad, 1938.

Reddy, V. Ramakrishna., *Economic History of Hyderabad State*, Gian Publishing House, Delhi, 1987.

Regani, Sarojini., *Nizam-British Relations, 1724–1857*, Swarajya Printing Works, Secunderabad, 1963.

———, 'The social background for the rise of the Andhra Mahasabha movement in Telangana', *Itihas*, vol. I, no. I, 1973.

Ribbentrop, B., *Forestry in British India,* 1900, Indus Publishing Company (Reprint), New Delhi, 1989.

Richards, J. F., 'The seventeenth century crisis in South India', *Modern Asian Studies*, vol. 24, Part 4, 1990.

———, 'The imperial crisis in the Deccan', *Journal of Asian Studies*, vol. 35, 1976, pp. 237–256.

Richards, J. F. and M. B. McAlpin, 'Cotton cultivating and land clearing in the Bombay Deccan and Karnatak, 1818–1920', in: R. P. Tucker and J. F. Richards (eds.), *Global Deforestation and the 19th century World Economy*, Durham, N. C., 1983.

Saberwal, Vasant K., 'Bureaucratic agendas and conservation policy in Himachal Pradesh, 1865–1994', *IESHR*, vol. 34, no. 4, 1997, pp. 465–498.

Saidulu, Rapolu., *Agrarian Conditions in Hyderabad State with Special Reference to Telangana: 1901–1948*, Unpublished PhD thesis, University of Hyderabad, Hyderabad, 1999.

Saldanha, I. M., 'The political ecology of traditional farming practices in Thana district, Maharashtra, India', *The Journal of Peasant Studies,* vol. 17(3), 1990.

———, 'Colonial forest regulations and collective resistance: 19th century Thana district', in: Richard Grove, V. Damodaran and S. Sangwan (eds.), *Nature and the Orient*, pp. 708–733.

Sangwan, Satpal., 'Making a popular debate: The Indian Forester and the emerging agenda of State forestry in India, 1875–1904', *IESHR*, April–June, 1999, vol. xxxvi, no. 2, pp. 187–238.

Saravana, Velayutham., 'Commercialisation of forests, environmental negligence and alienation of tribal rights in Madras Presidency: 1792–1882', *IESHR*, vol. 35, no. 2, 1998, pp. 125–146.

Sarkar, Sumit., 'Primitive rebellion and modern nationalism: A note on forest Satyagraha in the Non-Co-operation and Civil Disobedience movements', in: K. N. Panikkar (ed.), *National and Left Movements in India*, Oxford University Press, New Delhi, 1980.

Server, Adrian. (ed.), *Documents and Speeches on the Indian Princely States*, 2 vols., B. R. Publishing Corporation, Delhi, 1985.

Shiva, Vandana., *Staying Alive: Women, Ecology and Survival in India,* New Delhi, 1988. 'Ecology movements in India', *Alternatives*, vol. 11, 1986, pp. 255–273.

________, 'Common tragedy of the commons', in *EPW*, vol. 21, 1986, pp. 613–614.

Singh, Chhatrapati., *Common Property and Common Poverty: India's Forests, Forest Dwellers and the Law*, Oxford University Press, New Delhi, 1986.

Singh, St. Nihal., *The Nizam and the British Empire*, 1923.

Sirj-ul-Hassan, Syed, *The Castes and Tribes of HEH the Nizam's Dominions*, Bombay, 1920.

Sitaraman, Sarada., *Economic Survey of Sirsilk Ltd., Sirpur Kagaznagar*, MA, dissertation, Osmania University, Hyderabad, 1961.

Sivaramakrishnan, K., 'Colonialism and forestry in India: Imaging the past in present politics', In: *Comparative Studies in Society and History*, vol. 37, no. 1, January 1995, pp. 3–40.

________, 'A limited forest conservancy in South West Bengal, 1864–1912', *The Journal of Peasant Studies*, vol. 56, no. 1, Feb. 1997, pp. 75–112.

Stebbing, E. P., 'Forest conservancy in India', *Indian Forester*, vol. 19, no. 62, 1893.

Stebbing, E. P., *The Forests of India*, 3 vols., John Lane, London, 1922–1927. 'The forests of India and the development of the Indian Forest Department', in: *Indian Forester*, vol. 48, 1922, pp. 81–98.

________, 'The royal commission on agriculture in India and the Indian forests', In: *The Nineteenth Century*, vol. 105, 1929, pp. 327–330.

Sundarayya, P., 'Telangana peoples' armed struggle 1946–1951', *Social Scientist*, February 1973.

Sundarayya, P., *Telangana Peoples' Armed Struggle and its Lessons*, Communist Party of India (Marxist), Calcutta, 1972.

________, *Veera Telangana Viplava Poratam – Gunapathalu*, A. P. Committee CPI(M), Vijayawada, 1973.

Syeeduddin, M., *Some of the Common Flowering Plants of Hyderabad State, their Distribution, Economy and Medical Importance*, Hyderabad, 1935.

Tirtha, Swamy Ramananda, *Memoirs of Hyderabad Freedom Struggle*, Bombay, 1967.

Troup, R. S., *Indian Forest Utilisation*, Calcutta, 1913. 'Forestry in India', *Calcutta Review,* vol. 273, July 1913, pp. 305–312. *The Work of the Forest Department in India,* Calcutta, 1917.

Tucker R. P. and J. F. Richards (eds.), *Global Deforestation and the 19th century World Economy*, Duke University press, Durham, N. C., 1983.

Tucker, Richard., 'The Depletion of India's forests under British imperialism: Planters, foresters and peasants in Assam and Kerala', in: Donald Worster (ed.), *The Ends of the Earth – Perspectives on Modern Environmental History*, Cambridge University Press, 1988, pp. 118–140.

________, 'Forest management and imperial politics: Thana district, Bombay, 1823–1887', in *IESHR*, vol. 16, no. 3, 1979, pp. 273–300.

Turner, Michael., 'Economy, environment and environmental history: Case notes from agriculture', *Rural History*, vol. 8, no. 1, 1997, pp. 125–140.

Vaheed, Syed Abdul., A *Note on Post-War Forest Policy for the HEH the Nizam's Dominions*, Government Central Press, Hyderabad, 1944.

Vaikuntham Y., 'Growth of economic nationalism in Hyderabad state, A case study of Telangana – 1920–30' *Itihas*, vol. II, 1983.

Vartak, V. D. and Gadgil, Madhav., 'Sacred groves of India: A plea for continuous conservation', *Journal of Bombay Natural History Society*, vol. 2, no. 2, 1975.

Venkataramana, V., *Land Reforms in India: A Study of the Historical and Comparative aspects of Land Reforms in India with Special Reference to Composite Hyderabad State*, Unpublished PhD thesis, Osmania University, Hyderabad, 1957.

Visveswaraya, M., *The Flood of 1908 at Hyderabad: An Account of the Flood, Its Causes and Proposed Preventive Measures*, Venugopal Pillai and Sons, Hyderabad, 1909.

Voelcker, J. A., *Report on the Improvement of Indian Agriculture.*

Whitecombe, Elizabeth., 'The environmental costs of irrigation in British India: Waterlogging, salinity, Malaria', In: Ramachandra Guha and David Arnold (eds.), *Nature Culture Imperialism,* Oxford University Press, New Delhi, 1995.

————, *Agrarian Conditions in Northern India: The United Provinces under British Rule, 1860–1900*, Berkely, 1971.

Worster, Donald. (ed.), *The Ends of the Earth – Perspectives on Modern Environmental History*, Cambridge University Press, 1988. *Nature's Economy: A History of Ecological Ideas*, Cambridge, 1977.

Yazdani, G., *Early History of the Deccan*, 2 vols., New Delhi, 1983.

Index